SOILS AND FOUNDATIONS

ANDERSSON RINCON MOLINA

Copyright © 2021 Andersson Rincón Molina

All rights reserved.

ISBN: 9798731359962

DEDICATION

I dedicate this book to God, the Architect, Engineer and Builder of the universe and destiny, to my family, to my friends and colleagues, to my country, to the Cimarrón National Movement Association for the struggle of the rights and values of vulnerable communities, to the company where I laboro, to my teachers and to anyone who intervened directly and indirectly in my life contributing to my personal development , professional and social on the long road of my destiny, are those whom I must correspond to by contributing to their development and well-being through the exercise of my profession.

Table of contents

Table of contents vv .... *v*
List of figures ixviii .... *ix*
THANK YOU xvixiii .... *xvi*
INTRODUCTION 11 .... *1*
JUSTIFICATION 22 .... *2*
GENERAL OBJECTIVE 33 .... *3*
SPECIFIC OBJECTIVES 33 .... *3*
CHAPTER 1: 44 .... *4*
FOUNDATION 44 .... *4*
SURFACE FOUNDATIONS 55 .... *5*
CYCLOPEO FOUNDATION 66 .... *6*
ARMED CONCRETE FOUNDATIONS 66 .... *6*
6 6FOUNDATIONS *6*
FOUNDATION BY SHOES 77 .... *7*
FLOATING FOUNDATION 88 .... *8*
DEEP FOUNDATIONS 99 .... *9*
PILOT FOUNDATION 99 .... *9*

CHAPTER 2:
1011 .... *10*
FIELD AND LABORATORY TESTS MOST COMMONLY USED TO CALCULATE CARRYING CAPACITY
1111 .... *11*
FIELD TESTS
1111 .... *11*
The standard SPT penetration test:
1111 .... *11*
The vane cut resistance test:
1112 .... *11*
CBR "California Bearing Ratio":
1212 .... *12*
LABORATORY TESTS
1213 .... *12*
SOIL LABORATORY TESTS TO BE PERFORMED
1313 .... *13*
HUMIDITY CONTENT.
1414 .... *14*
ATTERBERG
17LIMITS 17 .... *17*
UNCONFINED COMPRESSION
2424 .... *24*
UNIT WEIGHT
3233 .... *32*
GRADATION
3940 .... *39*
CLASSIFSION SUCS
4647 .... *46*
CHAPTER 3:
4749 .... *47*
CLASSIFIES THICK SOIL SUCS
4749 .... *47*
EXAMPLE 1 FLOOR THICK
4850 .... *48*
EXAMPLE 2 THICK SOIL
5354 .... *53*
EXAMPLE 1 FINE FLOOR
5859 .... *58*

EXAMPLE 2 FINE FLOOR 6364 63
CHAPTER 4: 6869 68
DISTRIBUTION OF STRESSES OR PRESSURES ON THE GROUND UNDER A RIGID SHOE 6869 68
Boussinesq Theory 6869 68
FADUM 71 Table 73 71
Newmark 7577 method 75
Method two in one 7880 78
DEAD ...79 AND LIVE LOAD PER COLUMN BASED ON THE AFFERENT AREA. 79 81 79
INCREASED LOAD COMBINATIONS USING RESISTANCE METHOD 8183 81
LOAD COMBINATIONS 8183 81
EXAMPLE TRANSMISSION OF LOADS. 8385 83
LOAD DISTRIBUTION BY AFERENTE AREA 8689 86
LAST LOAD CAPACITY q_u 8991 89
PERMISSIBLE LOAD CAPACITY q_a 9092 90
SIZING AN ISOLATED SHOE AND A RUN. 9093 90
EXAMPLE ZAPATA RUN FOR WALL. 9093 90
ELASTIC THEORY 9093 90
NET REACTION CALCULATION 9194 91

CALCULATION OF THE PROFILE AND REINFORCEMENT BY MOMENT FLECTOR "HEIGHT" 92
EXAMPLE OF ISOLATED SHOE. 93
Punching cutter critical section to "d/2" of column (bidirectional cutter) 96
CHAPTER 5: TYPES OF FAILURES IN FOUNDATIONS 102
GENERAL CUT FAILURE 102
LOCAL COURT FAILURE 103
.... 105 PUNCH FAILURE
SOLUTIONS TO FOUNDATION FAILURES 106
TERZAGHI SOLUTION 106
SKEMPTON SOLUTION 108
CHAPTER 6: 108
TECHNICAL TERMS USED IN SOILS AND FOUNDATIONS 109
CONCLUSION 111
RECOMMENDATIONS 112
Bibliography 113
ABOUT AUTHOR 115

List of figures

Illustration 1. Constructive foundation process. 44
..4
Figure 2. Surface foundation. 55
..5
Figure 3. Cyclopeo foundation. 66
..6
Figure 4. Running foundation. 77
..7
Illustration 5. Foundation by shoes. 77
..7
Figure 6. Insulated shoe. 88
..8
Figure 7. Floating foundation. 99
..9
Figure 8. Deep foundation. 99
..9
Figure 9. Piloting foundation. 1010
..10
Figure 10. Standard SPT Penetration Test. Ambar 11Project 11
..11
Figure 11. Veleta Cut Resistance Test, Ambar Project. 1212
..12
Figure 12. CbR Test, Project Ambar. 1212
..12
Figure 13. Laboratory tests. 1414
..14
Figure 14. Balance and container. 1414
..14
Figure 15. Weight of the soil sample. 1515
..15
Figure 16. Sample of baked levada soil at 105oC. 1515
..15
Illustration 17. Weight of the dry soil sample, removed from the oven. 1616
..16
Figure 18. Ground states. 1717
..17

Illustration 19. Big house device. 18

Figure 20. Materials to use to determine liquid limit. 18

Figure 21. Sample kneading for liquid limit consistency. 19

Figure 22. Spoon is dropped from 10mm counting No of blows. ... 20

Figure 23. Hit number 20

Figure 24. Standardized graph. 21

Figure 25. Soil samples for plastic limit. 22

Figure 26. Materials for plastic limit testing. 22

Figure 27. Kneaded and molded ellipsoid in the sample. 23

Figure 28. Cylinder forming until cracks are seen and weighed in balance. .. 23

Figure 29. Sample drying on stove at 110o and heavy dry sample. ... 23

Figure 30. Limit plastic. 24

Figure 31. Incofined compression equipment. 26

Figure 32. Testing tools. 26

Figure 33. Sample dimensions, diameter height ratio. 27

Figure 34. Sample heavy. 28

Illustration 35. Initial stage of the trial. 29

Figure 36. Failure process. 29

Figure 37. Deformation/effort graph. 31

Figure 38. Equipment, tools and materials for unit weight testing. *3434*
....*34*
Figure 39. Soil sampling. *3536*
....*35*
Figure 40. Solid and liquid paraffin. *3637*
....*36*
Figure 41. Hydrostatic balance. *3637*
....*36*
Figure 42. Sieves for granulometric gradation. *3940*
....*39*
Figure 43. Sieves. *4142*
....*41*
Figure 44. Sample preparation. *4142*
....*41*
Figure 45. Granulometric curve, atterberg limits and SUCS AMBAR project classification. *4546*
....*45*
Figure 46. Plasticity graph. *4647*
....*46*
Figure 47. Results of laboratory tests. *4648*
....*46*
Figure 48.Probe Location. *4748*
....*47*
Illustration 49.Stratigraphic profile. *4749*
....*47*
Figure 50. Granulometric Curve Thick Floor Example 1. .... *4951*
....*49*
Illustration 51.Letter of Casagrande, "Letter of Plasticity" Thick Floor Example 1. .... *5052*
....*50*
Illustration 52.UsCS Soil Classification Thick Soil Example 1. .... *5152*
....*51*
Figure 53. Granulometric Curve Thick Floor Example 2. .... *5455*
....*54*
Illustration 54.Letter of Casagrande, "Letter of Plasticity" Thick Floor Example 2. .... *5556*
....*55*
Illustration 55.UsCS Soil Classification Thick Soil Example 2. .... *5657*
....*56*

Figure 56. Granulometric Curve Fine Floor Example 1. 5960
... 59
Figure 57. Letter from Casagrande, "Letter of Plasticity" Fine Floor Example 1. 6061
... 60
Illustration 58.UsCS Soil Classification Fine Floor Example 1. 6162
... 61
Figure 59. Granulometric Curve Fine Floor Example 2. 6465
... 64
Figure 60. Letter from Casagrande, "Letter of Plasticity" Fine Floor Example 2. 6566
... 65
Illustration 61.USCS Soil Classification Fine Floor Example 2. 6667
... 66
Figure 62. Theoria de Bousinesq. 6970
... 69
Figure 63.. Point Load Pressure Distribution, "Pressure Bulbs or Isobaras" 7071
... 70
Figure 64. Distribution of Efforts in Horizontal Plane. 7172
... 71
Figure 65. Distribution of Efforts in Vertical Plane. 7172
... 71
Figure 66.Graph to Determine FADUM Curves. 7375
... 73
Figure 67. Vertical Pressure Under a Uniform Load On a Circular Area. 7476
... 74
Figure 68. Newmark Method Pressure Distribution. 7578
... 75
Figure 69. Newmark Nomogram. 7779
... 77
Figure 70. Two-in-One Method Pressure Distribution. 7880
... 78
Illustration 71.Plan View of Structure to Be Analyzed. 8486
... 84
Figure 72. Structure Elevations to Analyze. 8587
... 85

Figure 73. Sections of Beams and Columns Structure to Analyze. 8587
....85
Figure 74. Structure Sn sn sn sn ss to analyze. 8688
....86
Figure 75. Distribution of Loads by Afferent Area of Structure Column to Analyze. 8789
....87
Figure 76. Cut and Plant Zapata Corrida. 9193
....91
Figure 77. Cumshot Shoe Scheme. 9295
....92
Figure 78. Cumshot Cumshot. 9396
....93
Figure 79. Insulated shoe. 9497
....94
Figure 80. Cut and Insulated Shoe Plant. 9497
....94
Figure 81. Punch cutter critical section to "d/2" of the column (bidirectional shear. 96 99
....96
Figure 82. General cut-off fault scheme: I-wedge of elastic state, II - active state zone, III - zone in passive state. 103103
.... 103
Figure 83. Settlement /Load by unit area q. 103103
.... 103
Figure 84. Local cutting failure scheme: I ' elastic state wedge, II ' active state zone. 104104
.... 104
Figure 85. Settlement /Load by unit area q. 104104
.... 104
Figure 86. Punch fault scheme: I s elastic state wedge. 105105
.... 105
Figure 87. Settlement /Load by unit area q. 106106
.... 106
Figure 88. Graph Load Capacity Factors, Terzaghi. 107108
.... 107

List of tables

Table 1. Granulometric Composition Thick Floor Example 1. 4950
.... 49
Table 2. USCS Unified Soil Classification System Thick Floor Example 1.
.... 5253
.... 52
Table 3. Granulometric Composition Thick Floor Example 2. 5455
.... 54
Table 4. USCS Unified Soil Classification System Thick Floor Example 2.
.... 5758
.... 57
Table 5. Granulometric Composition Fine Floor Example 1. 5960
.... 59
Table 6. USCS Unified Soil Classification System Fine Soil Example 1. 6263
.... 62
Table 7. Granulometric Composition Fine Floor Example 2. 6465
.... 64
Table 8. USCS Unified Soil Classification System Fine Soil Example 2. 6768
.... 67
Table 9. FADUM table. 7274
.... 72
Table 10. Table To Determine The Influence Coefficient "I". 7577
.... 75
Table 11. Mass of Materials. 8082
.... 80
Table 12. Uniformly Distributed Minimum Live Loads. 8183
.... 81
Table 13. Terzaghi 107107
.... 107 load capacity factors
Table 14. Skempton Load Capacity Values. 108109
.... 108

SOILS AND FOUNDATIONS

Thanks

I thank God, the Architect, Engineer and Builder of the Universe and the destiny, my Family, my country, the Cimarrón National Movement Association for the struggle of the rights and values of The Afro-descendant communities and the company where I lay down to give me the opportunity to pursue a professional career at such a prestigious university, for the trust they have placed in me and for their unconditional support.

SOILS AND FOUNDATIONS

Introduction

Through the development of this book it is intended to publicize knowledge and basic concepts regarding soils and foundations since soil and land are fundamental elements which influence construction projects in general and depending on their characteristics can be determined the type of foundation to be used as support of the building, soil plays a decisive role in supporting what is built, as usable material for embankments and/or fillers and even as building material in dams, dams or other works of common land in our Works.

SOILS AND FOUNDATIONS

Justification

Considering the need to know and evaluate the competencies that allow to identify concepts on Soils and Foundations, the development of this book covers the topic of soils and foundations where each component is resumed by studying it thoroughly and citing its testing methods, applicable standards, construction process, physical characteristics, which will strengthen the foundations of this knowledge for future performance in real situations.

SOILS AND FOUNDATIONS

GENERAL OBJECTIVE

To make known through this book the concepts that allow to identify the different types of foundations and understand the composition and behavior of the soils on which any construction project is supported whether building or civil works.

SPECIFIC OBJECTIVES

1. Identify the concepts of Soils and Foundations.

2. Differentiate the different types of foundations and their respective construction processes.

3. Conceptually understand the properties of the soils and the respective tests.

CHAPTER 1:

Cimentacion

The foundation **is the** set of structural elements whose mission is to transmit the loads of the building or elements supported to it to the **ground** by distributing them in such a way that they do not exceed their permissible pressure or produce zonal loads. Because the resistance of the soil is generally lower than that of the pillars or walls it will support, the area of contact between the ground and the foundation will be proportionally larger than the supported elements (except in very coherent rocky soils).

Foundation is important because it is the group of elements that support the superstructure. Particular attention must be paid as the stability of the construction depends to a large extent on the type of land.

Taking into account the above the foundation can be defined as the structural part of the building, responsible for transmitting the loads to the ground, this is the only element that we cannot choose, so the foundation will be made according to it. On the other hand, the terrain is not all at the same depth, another circumstance that influences the choice of the proper foundation.

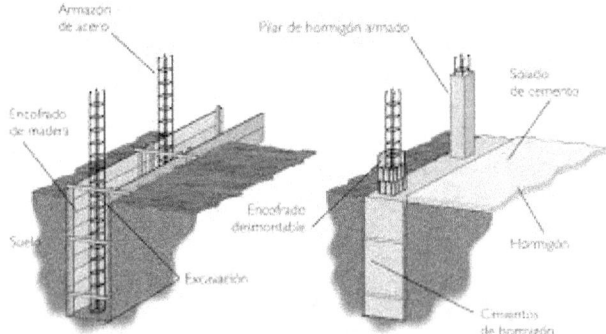

Illustration 1. Constructive foundation process.

The purpose of the foundation is to support structures by ensuring stability and avoiding damage to structural and non-structural materials.

Types of Foundations:
1. Surface foundations
2. Cyclopese foundation
3. Reinforced concrete foundations
4. Foundations run
5. Foundation by shoes
6. Floating foundation
7. Deep foundations
8. Pilot foundation

SUPERFICIAL FOUNDATIONS

Surface foundations are those that rest on the surface layers of the soil and are able to withstand the burden it receives from construction through base expansion. Stone is the most commonly used material in the construction of surface foundation, as long as it is resistant, solid and pore-free. However, reinforced concrete is an extraordinary building material and is always more recommended.

Illustration 2. Surface foundation.

SOILS AND FOUNDATIONS

CYCLOPEO FOUNDATION

In cohesive terrain where trench can be made with vertical parameters and without landslides, the foundation of cyclopese concrete is simple and economical.

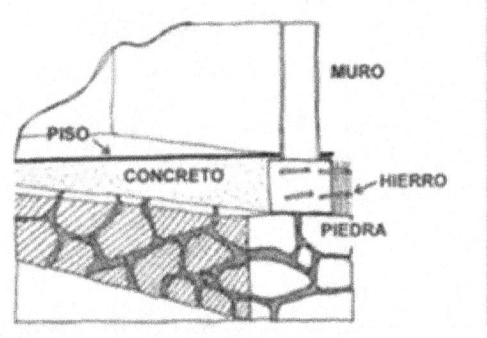

Illustration 3. Cyclopese foundation.

ARMED CONCRETE FOUNDATIONS

Reinforced concrete foundations are used in all terrains, although concrete is a heavy material, it has the advantage that relatively small sections are obtained in their calculation when compared to those obtained in the stone foundations.

FOUNDATIONS RUN

It is a type of concrete or reinforced concrete foundation that develops linearly at a depth and with a width that depends on the type of soil. It is used to properly transmit loads provided by bearing wall structures. It is also used to cement fence walls, gravity retaining walls, for high weight enclosures. Running foundations are not recommended when the soil is very soft.

Illustration 4. Running foundation.

FOUNDATION BY ZAPATAS

The shoes can be made of mass concrete or reinforced, with a square or rectangular plan, as well as foundation of vertical supports belonging to building structures, on homogeneous floors of sensibly horizontal stratigraphy.

Insulated foundation shoes shall be reinforced concrete for surface firms or mass for shallower firms, except those located on borders and medians. The depth of the support plane or choice of the firm shall be fixed according to the determinations of the geotechnical report, taking into account that the ground below the foundation is not altered. Previously to know what type of foundation we are going to use we need to know the type of terrain according to the geotechnical report.

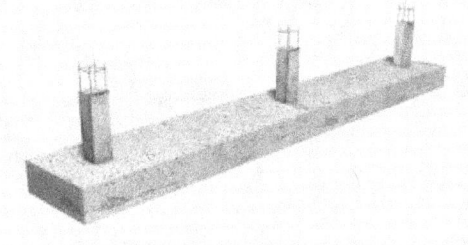

Illustration 5. Shoe foundation.

Types of shoes:
1.   Isolated shoes

2. Square insulated shoe
3. Rectangular insulated shoe
4. Off-center insulated shoe
5. Running shoes

ZAPATA AISLADA

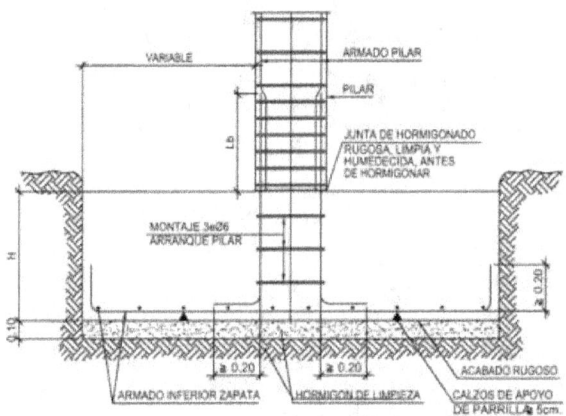

Figure 6. Insulated shoe.

FLOATING FOUNDATION

when the carrying capacity of the floor is very small and the weight of the building important, it may happen that the plot that we have does not have surface to house a sling that distributes the load; in such a case it is possible to build a foundation that floats on the ground.

Illustration 7. Floating foundation.

DEEP FOUNDATIONS

Deep foundations are responsible for transmitting the loads they receive from a construction to deeper resistant mantle. There are deep ones that transmit the load to the ground by pressure under its base, but they can also count on friction in the fuse; We classify them into:
1. Cylinders
2. Drawers

Illustration 8. Deep foundation.

PILOTING FOUNDATION

A pile is a support, usually made of reinforced concrete, of a large length in relation to its cross section, which can be swelled or built "in situ" in an open cavity in the ground. The piles are slender columns with the ability to withstand and transmit loads to stronger or rock strata, or by friction in the fuse. Usually, its diameter or side is not greater than 60 cms. It forms a construction system of deep foundation that we will call pilotingfoundation. Pilots are necessary when the surface layer or bearing floor is not able to withstand the weight of the building or when it is at great depth; also when the terrain is full of water and this makes digging difficult. The construction of stilts avoids expensive buildings and large volumes of foundation.

Illustration 9. Piloting foundation.

CHAPTER 2:

FIELD AND LABORATORY TESTS MOST COMMONLY USED TO CALCULATE CARRYING CAPACITY

Based on a soil study carried out prior to the construction of a project X located south of the city of Cali, it is possible to determine the tests carried out to calculate the carrying capacity which were field and laboratory and are cited below.:

FIELD TRIALS

The standard SPT penetration test:

It consisted of sinking the sampler or split spoon mentioned by dynamic action of a hammer weighing 140 Lbs falling from a height of 30 in, noting the number of strokes required to penetrate it into a length of 1 foot in the investigated stratum. We call this number "N" and it is the parameter of interest in the test. This allowed us to evaluate the consistency of cohesive soils, the compactness of granular soils and estimate the carrying capacity of the soil in general.

Figure 10. Standard SPT Penetration Test. Ambar Project

The vane cutting resistance test:

It consisted of measuring torque in N-m units, both in unchanged condition (Ti), and in towed condition (TR), parameters that allowed to determine the resistance to the cut not drained both in unaltered condition (Cui), and towed (CuR), correcting the values of Cu, for plasticity effects. The degree of soil sensitivity (S) corresponding to the relationship between unchanged cut resistance and towed cut resistance (S- Cui/ CuR) was also determined.

Figure 11. Veleta Cut Resistance Test, Ambar Project.

CBR "California Bearing Ratio":

To assess the support capacity of the sub-keeper that supports the pavements of the parkingers and internal route, an unchanged sample was taken in CBR mold, below the plant layer. You can see the sampling in CBR mold.

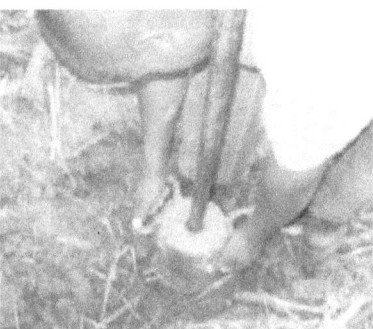

Figure 12. CbR Test, Project Ambar.

LABORATORY TESTS

In the laboratory, a test programme was submitted on the representative samples obtained from the surveys to determine their index properties such as:

- Humidity Content.
- Atterberg limits.
- Unconfined compression.
- Unit Weight.
- Gradation.

All of them were classified by the Unified Soil System (USCS).

SOIL LABORATORY TESTS TO BE PERFORMED

In the laboratory, representative samples obtained from the surveys undergo a test program to determine their index properties such as:

1. Humidity Content.
2. Atterberg limits.
3. Unconfined compression.
4. Unit Weight.
5. Gradation.

All of them were classified by the Unified Soil System (USCS).

The unaltered sample taken from the sub-engine was subjected to CBR testing in the laboratory in condition of both its natural and saturated moisture, i.e. after four (4) days of immersion in water, determining its percentage of expansion after saturation. The photograph shows the development of the test:

Figure 13. Laboratory tests.

The objective of each of the laboratory tests cited above is described below and its development also sets out its results.

HUMIDITY CONTENT.

It aims to determine the natural moisture of the soil by drying on a stove.

The following materials are used:

Figure 14. Balance and container.

1. Vessels and containers that support corrosion

The following procedure is performed:

An empty vessel (WR) is weighed, A soil sample is taken and its weight is placed in the vessel with the help of a balance (WT), The vessel with the soil sample is taken over it to an oven for a period of 24 hours at a temperature between 105oC and 115oC, The sample is taken from the oven and its dry mass (WS) is re-measured.

Figure 15. Weight of the soil sample.

Figure 16. Sample of baked levada soil at 105oC.

Figure 17. Weight of the dry soil sample, removed from the oven.

The parameters needed to determine the natural moisture of the soil are calculated from the known data.

Where:

WR: Mass of the container (Vase) - WR
WT: Total Sample Mass - WS + WW + WR
WTS: Total mass of the stove-dried sample - WS + WR
W: Natural soil moisture determined by stove drying (%)

$$W(\%) = \frac{(WT - WTS)}{(WTS - WR)} \times 100$$

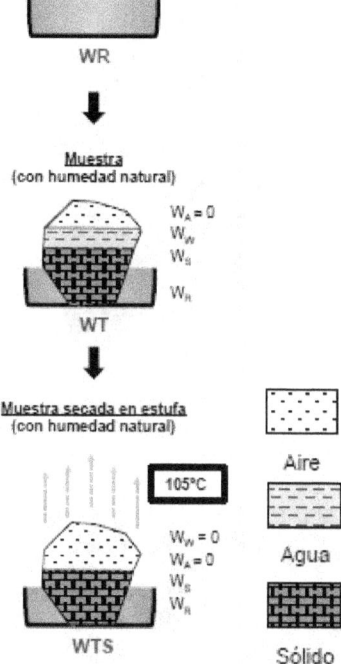

Figure 18. Ground states.

ATTERBERG LIMITS

It aims to determine the shrinkage limit, liquid limit and plastic limit, through these it is possible to get an idea of the characteristics of the type of soil under study.

Limit liquid.

It aims to determine the liquid limit of a soil by the method of the Casagrande apparatus.

Figure 19. Big house device.

The following materials are used:
2. Casagrande device, graters, spatulas and weights with lid.

Figure 20. Materials to be used to determine liquid limit.

The sample is prepared, kneading it to the humidity of the liquid limit approximately, on a smooth surface.

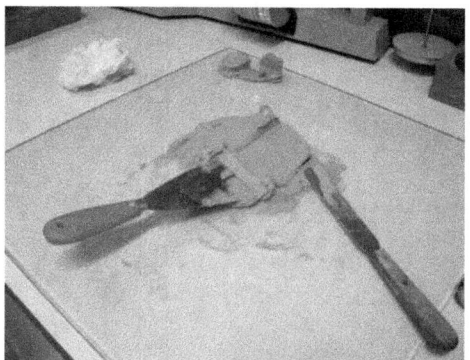

Figure 21. Sample kneading for liquid limit consistency.

The sample is left to be tested for 2 hours in wet chamber inside a plastic bag, then kneaded again, adding water if necessary at the end of this period, The sample is placed on the spoon of Casagrande, The groove is performed on the sample, The spoon is dropped from a height of 10 mm to a cadence of 2 strokes per second , The number of strokes (Ni) required is counted until the opening is closed along 13 mm, the number of strokes between 15 and 35 is counted, A sample of 10-15 gr is extracted from the junction area to obtain moisture (Wi), The same process is repeated for different humidities until a determination between 15 and 25 strokes and another between 25 and 35 strokes is obtained. The data is represented in the standardized chart and the humidity corresponding to 25 strokes corresponding to the Liquid Limit is obtained.

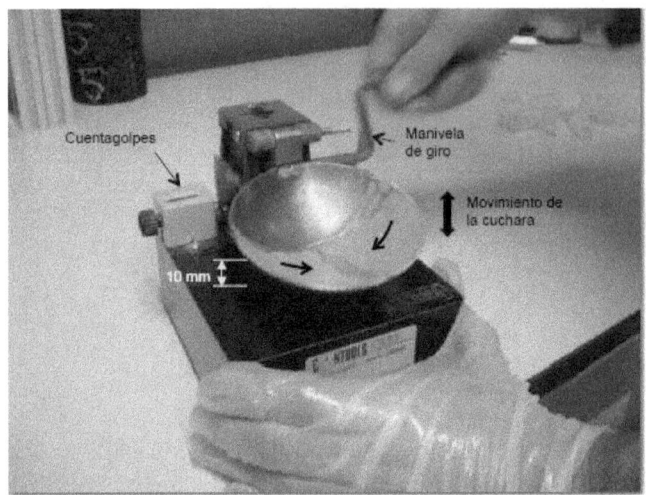

Figure 22. Spoon is dropped from 10mm counting No of blows.

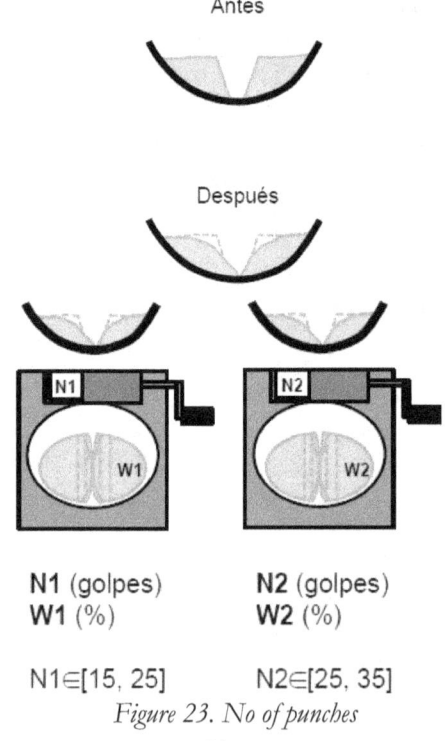

N1 (golpes) N2 (golpes)
W1 (%)   W2 (%)

N1∈[15, 25]   N2∈[25, 35]

Figure 23. No of punches

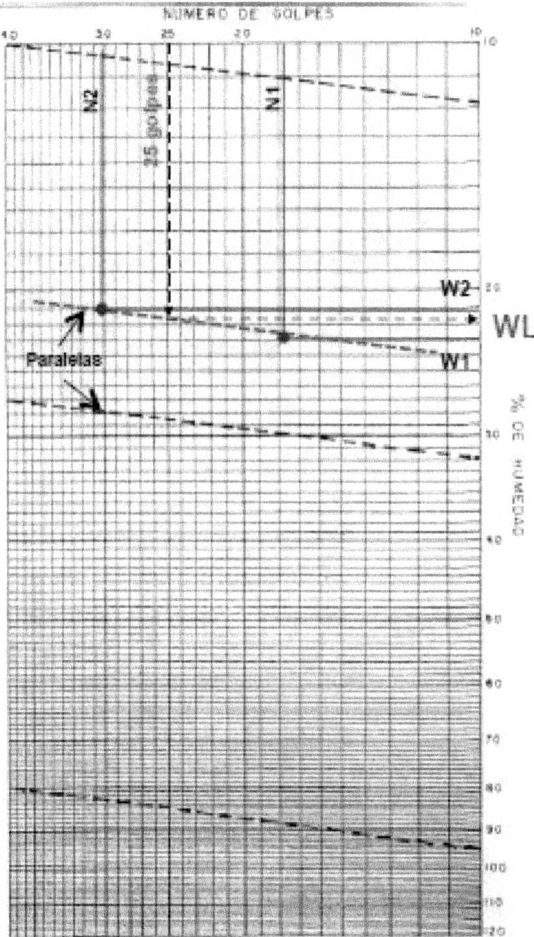

Figure 24. Standardized graph.

Limit plastic.
Aims to determine the plastic limit of a floor.

Figure 25. Soil samples for plastic limit.

The following materials are used:
3. Flexible blade spatula, smooth surface and weightssubstancing with lid.

Figure 26. Materials for plastic limit testing.

The sample is prepared, kneading it to the humidity of the plastic boundary, Half of the sample is molded in the form of an ellipsoid and rolled between the fingers, The cylinders are kneaded on a smooth surface, The cylinders are kneaded to a thickness of 3 mm in which cracks are appreciated, The cylinders are placed in the weights and weighed (WI) , The sample is dried on a stove at 110 oC for 24 hours, Sample heavy after 24 hours

(WII).

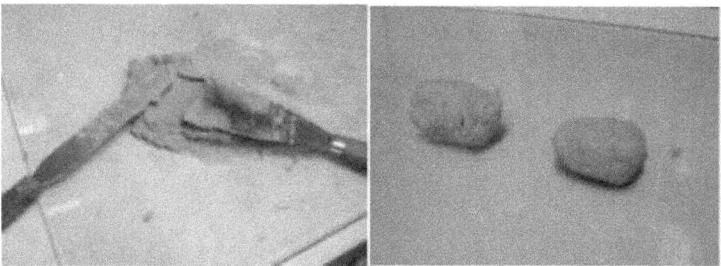

Figure 27. Kneaded and molded ellipsoid in the sample.

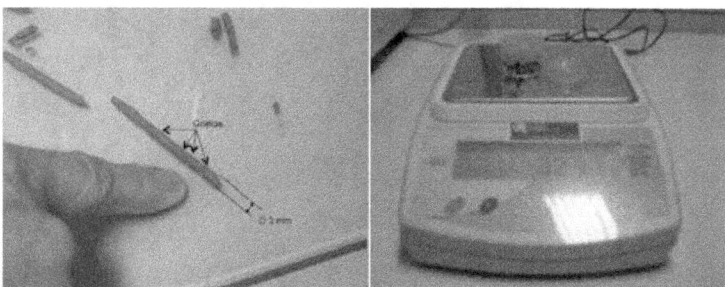

Figure 28. Cylinder forming until cracks are seen and weighed in balance.

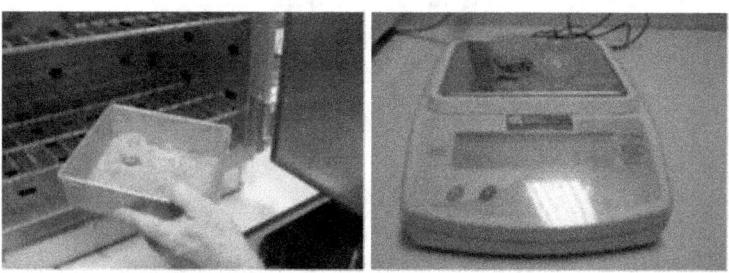

Figure 29. Sample drying on stove at 110o and heavy dry sample.

The plastic limit is the arithmetic mean of the humidity of the cylinders in both determinations expressed as a percentage.

Where:
W0: Mass of weightsseplaces empty
W1: Mass of the weightssubstantiances with floor cylinders

with a humidity equal to the plastic limit of the floor
W3: Mass of weightssubstancing with dry cylinders.

$$W_P(\%) = \frac{WII - WI}{WII - W0} \times 100$$

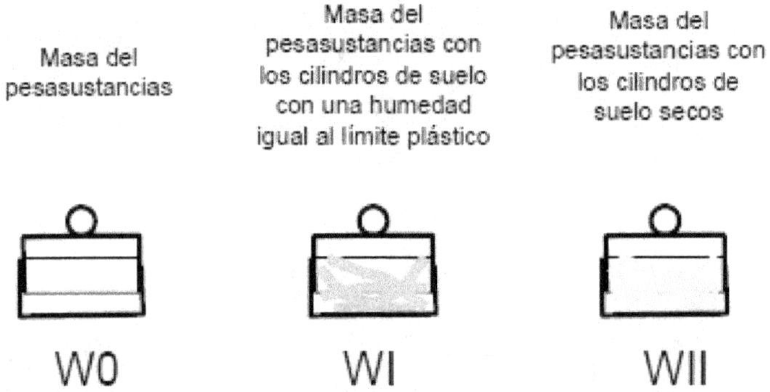

Masa del pesasustancias	Masa del pesasustancias con los cilindros de suelo con una humedad igual al límite plástico	Masa del pesasustancias con los cilindros de suelo secos
W0	WI	WII

Figure 30. Limit plastic.

INCONFINED COMPRESSION

It aims to determine and obtain data by means of which the following soil constants can be determined:
1. Sample Humidity Content
2. Unconfined Compression Resistance
3. Resistance to the Court
4. Graph Deformation / Effort

Team
1. Compression machine: compression instrument capable of applying normal forces at a constant speed and with a measuring device thereof, the accuracy of which varies depending on the type of material.
2. For soils whose unconfined compression resistance is estimated below 100 kPa (1kg/cm2), the compression machine must measure stress values with an accuracy of 1kPa (0.01 kg/cm2).
3. For soils whose unconfined compression resistance is

estimated above 100 kPa (1kg/cm2), the compression machine must measure stress values with an accuracy of 5kPa (0.05 kg/cm2).

4. Unaltered sample extractor: this type of test is carried out on preferably unchanged samples and it is of complete need to have a device that can extract samples from the sample tube and ensure that its condition is maintained.
5. Deformimeter: must be a caratula comparator, whose records provide an accuracy of 0.01 mm and a measurement length of at least 0.2 the total height of the test sample.
6. King's foot calibrator: calibrator with 0.01 mm precision in order to make accurate measurements to the test specimen.

1. Chronometer: time measurement instrument with 1s precision values. In this way, together with the data provided by the deformimeter it will be possible to measure the deformation rate of the specimen during the test.
2. Drying oven: with the capacity to maintain constant temperatures of $110 \pm 5oC$.
3. Sampling vessels: containers must have special characteristics such as: their material preferably aluminum, which supports high temperatures and is resistant to corrosion by contact with sample moisture. They must be marked with a code that facilitates their identification.
4. Gloves against high temperatures or tools to handle containers.
5. Precision balance: with 0.01 g precision, pre-calibrated.
6. Minor tools: minor tools such as spatulas, cleaning cloths, seguetas, among others.

Figure 31. Incofined compression equipment.

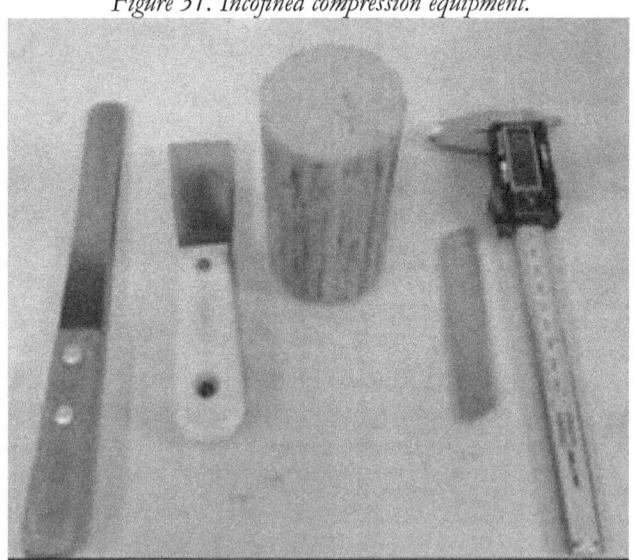

Figure 32. Testing tools.

Sample

The sample to be tested must have the following characteristics:
1. Minimum diameter of 30 mm.
2. Larger particle should be less than 0.1 times the diameter of the sample.
3. You must maintain a height diameter ratio of 2 to 3.

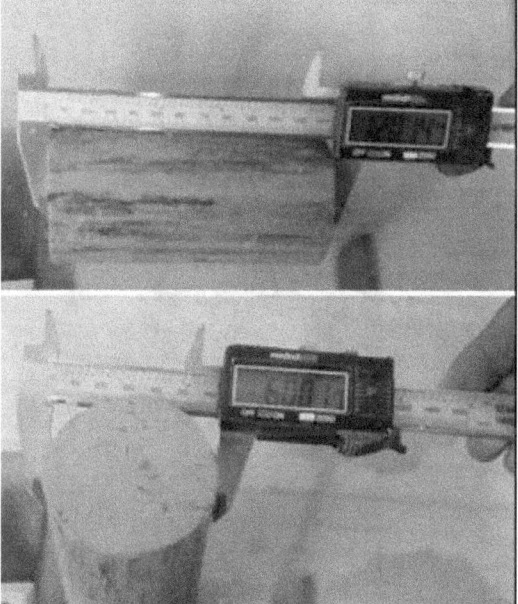

Figure 33. Sample dimensions, diameter height ratio.

Depending on the type of sample to be worked with (compacted, towed or unchanged) different conditions must be taken into account at the time of preparation and take into account the provisions of INV 103, 104, 105 as the case may be.

Procedure

The unconfined compression test on cohesive soils shall be terminated under the following conditions:
1. When the failure occurs, the cell that expresses the applied loads begins to show a drop in them and at that

point the sample is said to have failed.
2. The load is kept constant by four readings
3. Otherever, the test is said to be completed once a 20% strain has occurred, measured in axial deformation.

Before starting the unconfined compression test it is necessary to perform the calculation of the strain corresponding to 20 %, to know for sure when the test should finish if the failure has not occurred before.

Having clearly the above conditions, the test procedure is described:
1. Measure the physical and mass characteristics of the specimen. These correspond to weight, height and diameter.

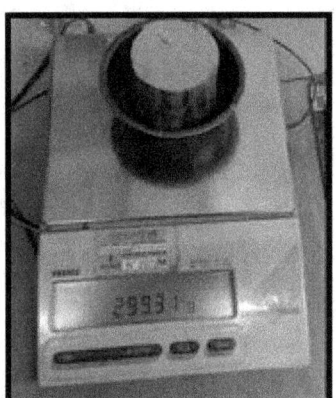

Figure 34. Sample heavy.

2. The specimen is placed at the base of the loading apparatus, making sure that the top side of the specimen matches the moving platen of the charging device, without any pre-deformation.

Figure 35. Initial stage of the trial.

3. Both the load reader, the deformometer and the timer are taken to zero reading.
4. The loading machine is operated and load readings are recorded and for the following deformation values when the deformometer has accuracy of 0.01 mm.

10, 25, 50, 75, 100, 150 and thereafter intervals of 50 deformation divisions until one of the above conditions is met.

1. Once the failure has occurred, graphically record the fault planes produced in the specimen.

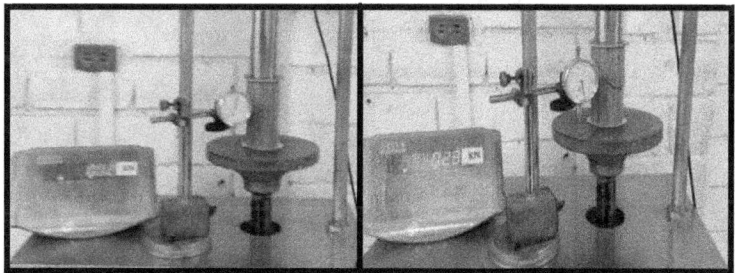

Figure 36. Failure process.

Calculations
- Area:

$$A = \pi * D^2 / 4$$

Where:
A: area of the sample face
D: Sample diameter
- Volume:

$$V = A * H$$

Where:
V: sample volume
H: sample height
- Dry Weight:

$$W_{seco} = W_{hum}/1+w$$

Where:
Wseco: dry sample weight
Whum: wet sample weight
w: Moisture content
- Wet Weight:

$$\gamma_{hum} = W_{hum}/V$$

Where:
Ahum: Wet Unit Weight
Whum: Wet Weight
V: Sample volume
- Unit Deformation:

$$\varepsilon = L/L_o * 100$$

Where:
ε : Strain
L: Changing the height of the specimen
Lo: Initial height of the specimen.
- Corrected area:

$$A_c = A_o/1 - \varepsilon 1/100$$

Where:
Ac: Corrected area

ε : Strain for each load
Ao: Initial area of the specimen cross-section
- Compression effort:
C sP/Ac

Where:
C: Compression effort
Q: Applied load
Ac: Corrected area
And whose greatest value presented will be the unconfined compression resistance of the specimen "qu".
- Cutting resistance:
Qu/2

Where:
:Cut resistance
qu: Unconfined compression resistance
- Graph Effort / Warp.

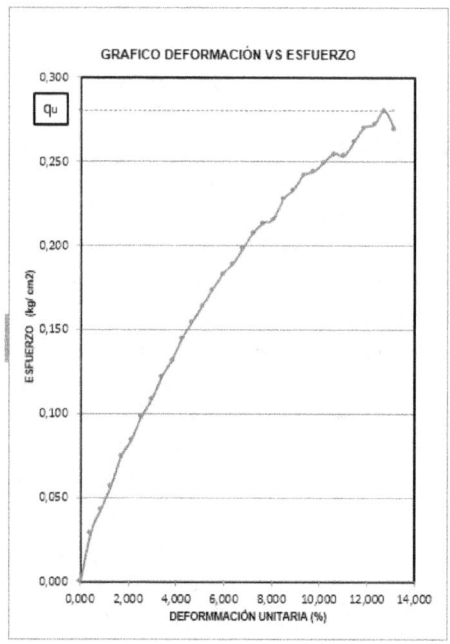

Figure 37. Deformation/effort graph.

Personal Protective Elements
2. Steel-tipped boots
3. Safety goggles
4. Safety gloves
5. Overol or work gown.

Report
In the report you must submit the following information:
1. Project name, location, polling number, stack or trench, sample number, and depth.
2. Soil Description and Classification
3. Initial Humidity Content
4. Equipment used during the test.
5. Test condition (saturated or natural moisture)
6. Sample type (unaltered, towed or compacted)
7. Fault diagram. Front and rear view.
8. Compression resistance, cut resistance, and maximum deformation values.
9. Graph Deformation vs Effort.

UNIT WEIGHT

The unit weight of a soil can be defined as the mass of a unit volume of soil, in which the volume includes the volume of the individual particles and the volume of voids between particles, whether these voids are filled with water for which they would be saturated unit weight or that are dry for a dry unit weight. The value of the unit weight of the soil in addition to varying by the amount of water that the soil has (dry, wet or saturated condition), will also depend on compaction and consolidation conditions that it presents.

A saturated unit weight is defined as the weight of the saturated soil mass per unit volume, where the voids are filled with water.

A wet unit weight is defined as the weight of the soil mass per unit volume, where soil voids contain as much water as air.

A dry unit weight is defined as the weight of dry soil mass per unit volume, where vacuums do not contain water.

The purpose of this test method is to obtain data by which the unit weight of the soil can be determined, which will be used as input in subsequent calculations for the determination of different soil properties.

Team

Regular geometric figure tilling method

1. Sample labrador: equipment to give the sample a regular, normally cylindrical shape. You must ensure the clamping of the sample and your blade or thread must have enough edge to leave the surfaces as finished as possible.
2. King's foot calibrator: considering that the test depends to a large extent on the accuracy of the volume being calculated, the diameter and height dimensions should be taken as accurately as possible. It is recommended that the digital cell measurements calibrator, to avoid operator errors.
3. Drying oven: with the capacity to maintain constant temperatures of $110 \pm 5oC$.
4. Precision balance: with 0.01 g accuracy, pre-calibrated
5. Knives and miscellaneous tools that may be needed.
1. Water immersion method using paraffin
2. Knife for labrare thin thread
3. Precision balance: with 0.01 g accuracy, pre-calibrated
4. Diluted paraffin. Your density must be known.
5. Stove
6. Volumetric vessel

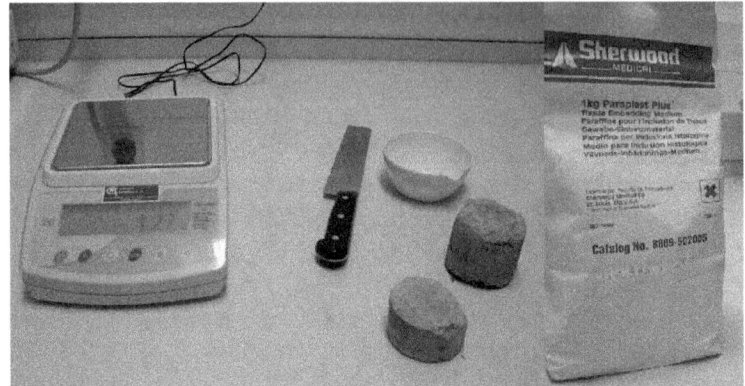

Figure 38. Equipment, tools and materials for unit weight testing.

Other Factors
1. Temperature: To keep the sample humidity unchanged, the sites where the tests are carried out must not have temperature variations greater than ±4oC, nor should they have direct contact with sunlight.
2. Sampling and Storage: Sampling a soil is the pre-analysis and property determination stage. It is probably the most important phase for obtaining analytical data that can be considered safe and to be able to make a truthful opinion on the soil under analysis.
3. Its storage must ensure that the sample retains its natural moisture and that there are no events of volume changes in it. Avoid hitting samples and producing cracks in samples

Procedure
Regular geometric figure tilling method
1. From the field sample, which must be unchanged and extracted by shelby, the sample is cut with a geometry of approximately 5x5x11 cm.
2. Once the above condition is had, the sample is carried to the labrador apparatus is carried to a cylindrical shape whose diameter is 36mm.

3. Cut the ends of the sample perpendicularly as accurately as possible. It is recommended that the sample height be 10 cm.
4. The measurements of the sample are carried out, corresponding to diameter and height. It is recommended to make 3 diameter measurements: one upper, one central and finally one higher.
5. Finally, the sample is carried to the balance and its weight is determined.
6. If the material resulting from the tilling is sufficient, it can be collected to determine moisture content.

Figure 39. Soil sampling.

Water immersion method using paraffin

1. Soak the sample until you get a side cube at 3.0 and 4.0 cm.
2. Bring the towed sample to the balance and record its weight.
3. Taking paraffin to a high-temperature resistant bottle is a stove to dilute it until a completely fluid state is present.
4. The sample should be tied with a cord or thread and immersed in the paraffin bottle. The immersion bath should ensure that the layer is thin and uniform.
5. Once the paraffin has cooled and dried, the sample is taken back to the balance and its weight is recorded, all

SOILS AND FOUNDATIONS

without removing the mooring with the yarn that has been made previously.
6. Put water in the glass with millimeter division, write down its initial volume and immerse the paraffin sample in it. Write down the volume that registers the glass once the sample has been entered into it.

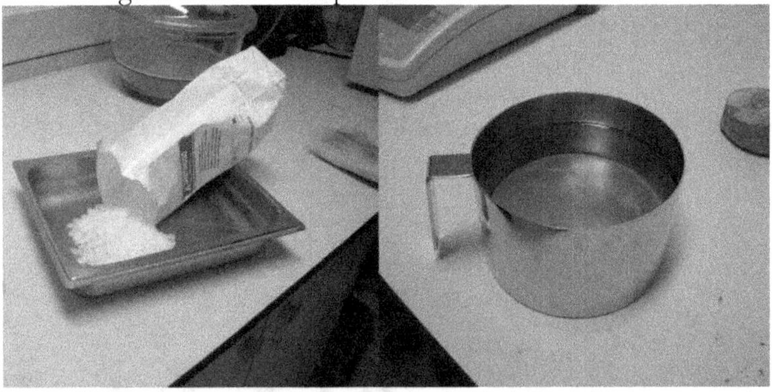

Figure 40. Solid and liquid paraffin.

Figure 41. Hydrostatic balance.

Calculations:
Regular geometric figure tilling method

SOILS AND FOUNDATIONS

- Area:
$Ap = As+4Ac+Ai/6$
Where:
As: determined area with top diameter.
Ac: determined area with central diameter.
Ac: determined area with lower diameter.
Ap: average area.

- Volume:
$V = Ap*Hm$
Where:
V: sample volume
H: sample height
Ap: average area

- Wet Unit Weight:
$\gamma hum = Wm/V$
Where:
Ahum: Wet Unit Weight
Wm: Wet Weight
V: Sample volume

- Dry Weight:
$Wseco = Whum/1+w$
Where:
Wseco: dry sample weight
Whum: wet sample weight
w: Moisture content

- Dry Unit Weight:
$Oseco = Wseco/V$
Where:
Ahum: Dry Unit Weight
Wseco: Dry weight
V: Sample volume

Water immersion method using paraffin
- Paraffin volume :
$Vp = Pmp-Pmi / p$

Where:
Vp: Paraffin volume
Pmp: paraffin sample weight
Pmi: initial sample weight
P: density of paraffin

- Sample volume :
Vm-Vad-Vp
Where:
Vm: Sample volume
Vad: Volume of evicted water
Vp: Paraffin volume

- Wet Unit Weight:
γ hum?Wm/Vm
Where:
Ahum: Wet Unit Weight
Wm: Wet Weight
Vm: Sample volume

If you want to know the dry unit weight, you proceed to perform the calculations described for the regular geometry sample method.

Personal Protective Elements
1. Steel-tipped boots
2. Safety goggles
3. Safety gloves
4. Overol or work gown.

Report
In the report you must submit the following information:
1. Project name, location, polling number, stack or trench, sample number, and depth.
2. Soil Description and Classification
3. Initial Humidity Content
4. Equipment used during the test.
5. Test condition (saturated or natural moisture)
6. Sample type (unaltered, towed or compacted)
7. Wet and dry unit weight value.

Gradation

Sift granulometry

Granulometric analysis of floors by sieving.

Figure 42. Sieves for granulometric gradation.

Granulometric analysis consists of separating soil particles by size ranges, making use of meshes or sieves with square openings.

This test method is intended to obtain data by which the following soil constants can be determined:
1. Uniformity coefficient
2. Curvature coefficient
3. Percentage of tax
4. Percentage of sands
5. Percentage of fine
6. Soil classification according to SUCS
7. Granulometric curve

Team
1. Sieve set: The sieves referred to in Table 3 are set as

necessary in order to achieve key points in the realization of the granulometric curve.

It is necessary that the set of sieves has a background and top cover. These must be in good condition and free of bumps on their edges that cause their bonding and separation with the sieves to occur by sudden movements; this in order to avoid unexpected material drops.

2. Stirring device: a mechanical stirrer that allows the sieving process to be performed.

In case of no stirrer, the process can be carried out manually, taking care that material leakage occurs due to movements made by the operator. The stirring intervals should last long enough, thus ensuring that the meshes allow all particles smaller than their hole size to pass through.

3. Drying oven: with the capacity to maintain constant temperatures of 110 ± 5oC.
4. Sampling vessels: containers must have special characteristics such as: their material preferably aluminum, which supports high temperatures and is resistant to corrosion by contact with sample moisture. They must be marked with a code that facilitates their identification.
5. Scale #1: with precision of 0.01 g, pre-calibrated.
6. Scale #2: with precision of 0.1 g , pre-calibrated.
7. Wire brush
8. Thin-haired brush.

Figure 43. Sieves.

The preparation of the sample is carried out by the quartet method and once it is carried out the sample is separated into thin and thick by means of the washing method.

Figure 44. Sample preparation.

Procedure

To. Floor retained in sieve No. 10 (2mm)

1. The series of sieves to be used is prepared. Taking into account Table 3, sieves will be taken from No. 10 to the 3" sieve.
2. The material that has been defined and prepared for the test was deposited from the top of the sieve tower and the top cover is adjusted.
3. Either mechanically or manually the sieving process is carried out. If the sieving process is manual, it is

recommended to perform movements that produce sufficient vibration and the displacement of particles from one place to another through the meshes of the sieves.

The passage of each of the soil particles through the sieve holes should be given freely and at no time should the operator force the particles to pass any of the sieve meshes.

4. With the help of measuring vessels and a 0.01 g precision balance, the weights of the materials retained in each mesh shall be carried out.
5. This operation will be carried out by first dismantling the larger diameter sieves, depositing their contents in containers placed on the balance and recording the weight of the material.

B. Floor passes sieve No. 10 (2mm)

According to INV-123 the proportions for typically sandy soils should be greater by a ratio of 100% compared to typically clayey or slimy soils.

1. In a cylindrical container of known capacity, approximately 250 ml of water is added and the soil is mixed into it. In addition, 125 ml of hexametaphosphate can be added which will act as a dispersant agent.

The containers to be used should preferably be made of aluminum or porcelain, as at a later stage they will be put in the oven to carry out the drying process.

2. Once the mixture occurs and you are sure that surface wetting of the particles occurs, the mixture should be left to rest for a period of at least 12 hours.
3. After the saturation period, the mixture should be deposited on the sieve No. 200, performing the washing process that was explained in procedure a.
4. Once the sample has been washed, it is deposited in a new container. Adhesion of the particles to the sieve mesh shall be presented, so distilled water should be used to ensure the transfer of the entire sample.
5. The sample is carried into the oven for drying at a temperature of 110oC ±5, until a constant mass is

6. Once the material is at room temperature, the series of sieves will be prepared from No. 20 to No. 200 and available from the top.
7. Either mechanically or manually the sieving process is carried out. If the sieving process is manual, it is recommended to perform movements that produce sufficient vibration and the displacement of particles from one place to another through the meshes of the sieves.

The passage of each of the soil particles through the sieve holes must be given freely and at no time should the operator force the particles to pass any of the sieve meshes. If it is necessary to use the brush, this operation should be performed from the outside of the mesh, in order to return the particle, but not force it to pass.

8. With the help of measuring vessels and a 0.01 g precision balance, the weights of the materials retained in each mesh shall be carried out.
9. This operation shall be carried out by first dismantling the larger diameter sieves, depositing their contents in containers placed on the balance and recording the weight of the material.

Calculations
- Percentage Retained:
% Retained- Mr/MT*100
Where:
Mr: Mass retained in the sieve
MT: Total mass

- Accumulated Retained Percentage:
% Retained Acormulado-Sumof percentages greater than or equal to

- Passing Percentage:
Passing %100% retained encomulated

- Uniformity Coefficient:
$C_U = D_{60}/D_{10}$
Where:
D60 : Sizes of soil particles in millimeters, which in the graph of the granulometric composition correspond to 60 %.

D10: Sizes of soil particles in millimeters, which in the graph of the granulometric composition correspond to 10 %.

- Curvature Coefficient:
$C_U = (D_{30})^2/(D_{10}*D_{60})$
Where:
D60 : Sizes of soil particles in millimeters, which in the graph of the granulometric composition correspond to 60 %.

D10: Sizes of soil particles in millimeters, which in the graph of the granulometric composition correspond to 10 %.

D30: Sizes of soil particles in millimeters, which in the graph of the granulometric composition correspond to 30 %.

The following criterion is handled for Cu and Cc values:

Cu>4 y Cc entre 1 y 3	GW
Cu >6 y Cc entre 1 y 3	SW
Si no cumple GP o SP	

GW	GRAVA BIEN GRADADA
SW	ARENA BIEN GRADADA
GP	GRAVA MAL GRADADA
SP	ARENA MAL GRADADA

- Granulometric Curve

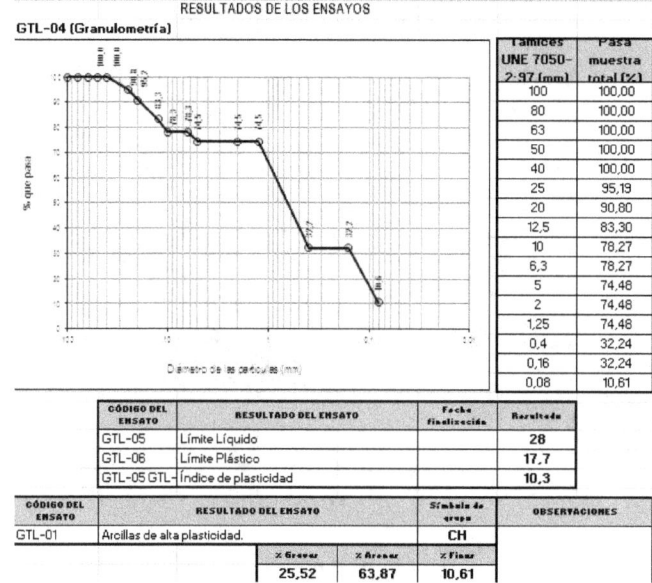

Figure 45. Granulometric curve, atterberg limits and SUCS AMBAR project classification.

Personal Protective Elements
1. Steel-tipped boots
2. Safety goggles
3. Safety gloves
4. Overol or work gown.

Report
In the report you must submit the following information:
1. Project name, location, polling number, stack or trench, sample number, and depth.
2. Soil Description and Classification
3. Equipment used during the test.
4. Test condition (saturated or natural moisture)
5. Sample weight passes sieve No. 200 and Sample weight Retained in Sieve No. 200
6. Values of D10, D30, and D60.
7. Percentages of Gravas, Arenas and Finos.

SOILS AND FOUNDATIONS

8. Soil classification according to "SUCS".
9. Granulometric Curve.

CLASSIFSION SUCS

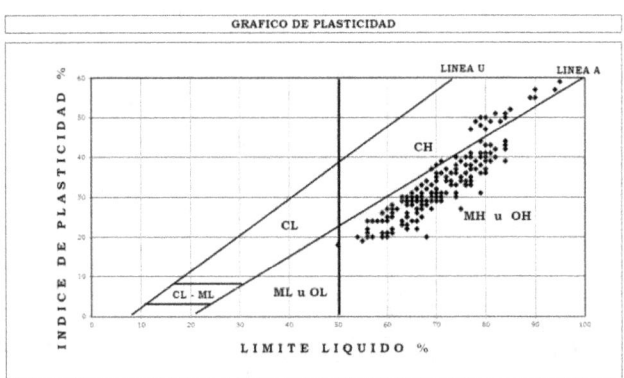

Figure 46. Plasticity graph.

In accordance with the plasticity graph it can be observed that the soil in which the Ambar project is developed contains high plasticity elements such as High Plasticity Limos (MH), High Plasticity Organics (OH), and High Plasticity Clays (CH).

Figure 47. Results of laboratory tests.

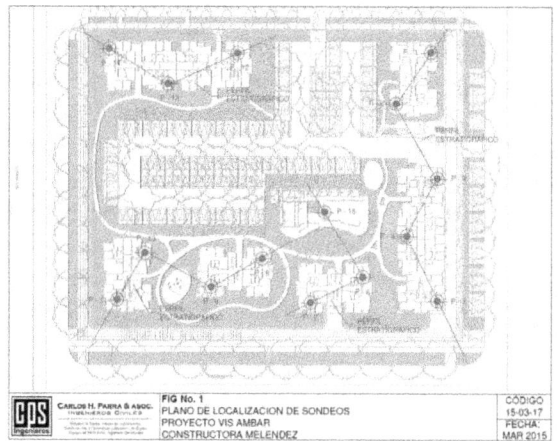

Figure 48. Location of probes.

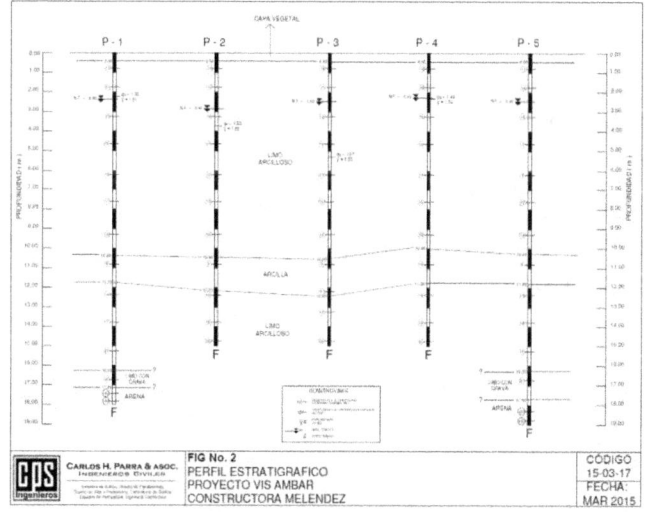

Figure 49. Stratigraphic profile.

CHAPTER 3:

CLASSIFIES SUCS OF THICK SOILS

In accordance with the SUCS Unified Soil Classification

System, the thick soils differ from the thin ones by sieving or screening the material by mesh No. 200, the thick soils correspond to those retained in that mesh, and the fine ones that pass it, and thus a soil is considered thick if more than 50% of the particles of it are retained in the mesh No. 200 , and fine if more than 50% of its particles are smaller than that mesh.

In the coarse soils are the gravel (G) and sands (S), so that a soil belongs to the group of gravel (G) if more than half of the thick fraction is retained by mesh No. 4, and belongs to the group of sands (S) otherwise, i.e. the gravel (G) are between the meshes 3" and No. 4 and the sands between the meshes No. 4 and No. 200.

Here are 2 examples of thick soil classification:

EXAMPLE 1 THICK SOIL

It has a soil to which the corresponding tests have been performed such as granulometry and atterberg limits where the following results were obtained:

Atterberg limits.
LL: Liquid limit: 27%
LP: Plastic limit: 8%
IP: Plasticity Index - LL – LP: 19%

Granulometry.
It ranges from the mesh of the sieve No. 3/4" to the mesh No. 200 and has a percentage of material that passes the mesh No. 200 by 27% below is the table of granulometric composition of the material and graph of the granulometric curve of the soil cited.

SOILS AND FOUNDATIONS

Tamiz (mm)	Malla	Pasa (%)	Pasante (%)	Retenido acumulado (%)	Retenido parcial (%)
101,6	4"	100,00	100,00	0,00	0,00
76,2	3"	100,00	100,00	0,00	0,00
50,8	2"	100,00	100,00	0,00	0,00
38,1	1 1/2"	100,00	100,00	0,00	0,00
25,4	1"	100,00	100,00	0,00	0,00
19,1	3/4"	90,00	90,00	10,00	10,00
9,52	3/8"	85,00	85,00	5,00	5,00
4,76	No.4	60,00	60,00	25,00	25,00
2,00	No.10	54,00	54,00	6,00	6,00
0,84	No.20	50,00	50,00	4,00	4,00
0,42	No.40	36,00	36,00	14,00	14,00
0,25	No.60	33,00	33,00	3,00	3,00
0,149	No.100	29,00	29,00	4,00	4,00
0,074	No.200	27,00	27,00	2,00	2,00
	FONDO	0,00	0,00	27,00	27,00
			TOTAL	100,00	100,00

Límite líquido LL	27,00	%
Límite plastico LP	8,00	%
Indice plasticidad IP	19,00	%

Pasa tamiz N° 4 (4,76mm):	60,00	%
Pasa tamiz N° 200 (0,074mm):	27,00	%
D_{60}:	4,76	mm
D_{30}:	0,17	mm
D_{10} (diámetro efectivo):	0,03	mm
Coeficiente de uniformidad (Cu):	173,68	
Grado de curvatura (Cc):	0,23	

Table 1. Granulometric Composition Thick Floor Example 1.

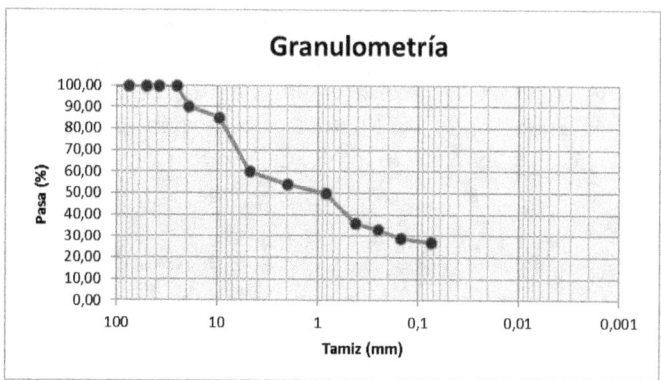

Figure 50. Granulometric Curve Thick Floor Example 1.

Taking into account the results of the atterberg boundaries it can be observed that the indicator point in the plasticity chart "Letter of Casagrande" is above line A and on the left side of line

B therefore this indicates that the ground is in
the area of clays of low cl plasticity, below is the graph of the plasticity chart.

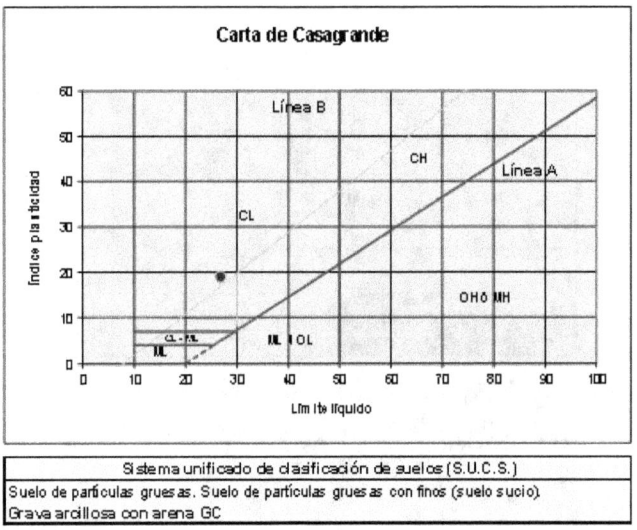

Figure 51. Letter from Casagrande, "Letter of Plasticity" Thick Floor Example 1.

In line with the results obtained it can be determined that through the USCS Unified Soil Classification System, the soil belongs to the Coarse **Grain**Soils, Gravas, Fine Gravas, Clay Gravel, Grave-Sand-Clay Mixtures and its **symbol is GC,**below is the graph and table of theUSCS Unified Soil Classification System.

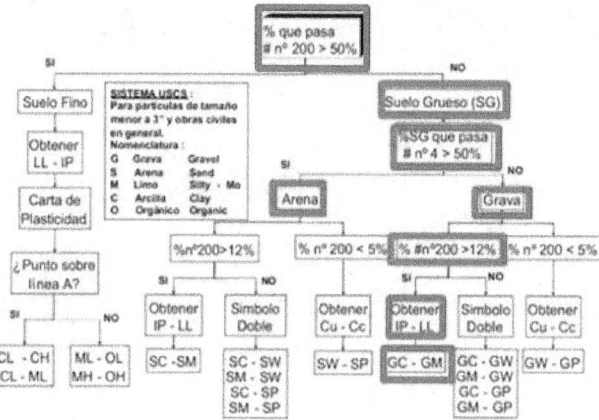

Figure 52. SOIL Classification USCS Thick Floor Example 1.

SISTEMA DE CLASIFICACIÓN DE SUELOS UNIFICADO "U.S.C.S."

DIVISIONES PRINCIPALES			Símbolos del grupo	NOMBRES TÍPICOS	IDENTIFICACIÓN DE LABORATORIO		
SUELOS DE GRANO GRUESO Más de la mitad de la fracción gruesa es retenida por el tamiz número 4 (4,76 mm)	GRAVAS Más de la mitad de la fracción gruesa es retenida por el tamiz número 4 (4,76 mm)	Gravas limpias (sin o con pocos finos)	GW	Gravas, bien graduadas, mezclas grava-arena, pocos finos o sin finos.	Determinar porcentaje de grava y arena en la curva granulométrica. Según el porcentaje de finos (fracción inferior al tamiz número 200). Los suelos de grano grueso se clasifican como sigue: <5%->GW,GP,SW,SP. >12%->GM,GC,SM,SC. 5 al 12%->casos límite que requieren usar doble símbolo.	$Cu=D_{60}/D_{10}>4$ $Cc=(D30)^2/D_{10}xD_{60}$ entre 1 y 3	
			GP	Gravas mal graduadas, mezclas grava-arena, pocos finos o sin finos.		No cumplen con las especificaciones de granulometría para GW.	
		Gravas con finos (apreciable cantidad de finos)	GM	Gravas limosas, mezclas grava-arena-limo.		Límites de Atterberg debajo de la línea A o IP<4.	Encima de línea A con IP entre 4 y 7 son casos límite que requieren doble símbolo.
			GC	Gravas arcillosa, mezclas grava-aren-arcilla.		Límites de Atterberg sobre la línea A con IP>7.	
	ARENAS Más de la mitad de la fracción gruesa pasa por el tamiz número 4 (4,76 mm)	Arenas limpias (pocos o sin finos)	SW	Arenas bien graduadas, arenas con grava, pocos finos o sin finos.		$Cu=D_{60}/D_{10}>6$ $Cc=(D30)^2/D_{10}xD_{60}$ entre 1 y 3	
			SP	Arenas mal graduadas, arenas con grava, pocos finos o sin finos.		Cuando no se cumplen simultáneamente las condiciones para SW.	
		Arenas con finos (apreciable cantidad de finos)	SM	Arenas limosas, mezclas de arena y limo.		Límites de Atterberg debajo de la línea A o IP<4.	Los límites situados en la zona rayada con IP entre 4 y 7 son casos intermedios que precisan
			SC	Arenas arcillosas, mezclas arena-arcilla.		Límites de Atterberg sobre la línea A con IP>7.	
SUELOS DE GRANO FINO Más de la mitad del material pasa por el tamiz número 200	Limos y arcillas: Limite líquido menor de 50		ML	Limos inorgánicos y arenas muy finas, limos limpios, arenas finas, limosas o arcillosa, o limos arcillosos con ligera plasticidad.	Ábaco de Casagrande		
			CL	Arcillas inorgánicas de plasticidad baja a media, arcillas con grava, arcillas arenosas, arcillas limosas.			
			OL	Limos orgánicos y arcillas orgánicas limosas de baja plasticidad.			
	Limos y arcillas: Limite líquido mayor de 50		MH	Limos inorgánicos, suelos arenosos finos o limosos con mica o diatomeas, limos elásticos.			
			CH	Arcillas inorgánicas de plasticidad alta.			
			OH	Arcillas orgánicas de plasticidad media a elevada; limos orgánicos.			
Suelos muy orgánicos			PT	Turba y otros suelos de alto contenido orgánico.			

Table 2. USCS Unified Soil Classification System Thick Floor Example 1.

The soil is thick as the percentage passed by the sieve No. 200 is less than 50% and belongs to the Arenas group because the percentage passed by mesh No. 4 is greater than 50%, en Grava because more than 50% of the thick fraction is retained in sieve No. 4, it is Clay gravel GC because the percentage passed by the sieve No. 200 is greater than 12% , through the above information we can obtain the following data.

    a. Amount of thick soil: 73%

b. Amount of fine soil: 27%
c. Amount of Gravel: 40%
d. Amount of sand: 33%
e. Amount of coarse gravel: 15%
f. Average amount of sand: 14%
g. Amount of fine sand: 9%
h. Soil classification: GC Clay gravel, gravel-sand-clay mixtures.
i. Plasticity type: CL Low plasticity clays.

EXAMPLE 2 THICK SOIL

It has a soil to which the corresponding tests have been performed such as granulometry and atterberg limits where the following results were obtained:

Atterberg limits.
LL: Liquid limit: 27%
LP: Plastic limit: 21%
IP: Plasticity Index - LL – LP: 6%

Granulometry.
It comprises from the mesh of the sieve 1" to the mesh No. 200 and has a percentage of material that passes the mesh No. 200 by 4% below is shown the table of granulometric composition of the material and graph of the granulometric curve of the soil cited.

Tamiz (mm)	Malla	Pasa (%)	Pasante (%)	Retenido acumulado (%)	Retenido parcial (%)
101,6	4"	100,00	100,00	0,00	0,00
76,2	3"	100,00	100,00	0,00	0,00
50,8	2"	100,00	100,00	0,00	0,00
38,1	1 1/2"	100,00	100,00	0,00	0,00
25,4	1"	96,00	96,00	4,00	4,00
19,1	3/4"	93,00	93,00	3,00	3,00
9,52	3/8"	86,00	86,00	7,00	7,00
4,76	No.4	70,00	70,00	16,00	16,00
2,00	No.10	49,00	49,00	21,00	21,00
0,84	No.20	45,00	45,00	4,00	4,00
0,42	No.40	17,00	17,00	28,00	28,00
0,25	No.60	10,00	10,00	7,00	7,00
0,149	No.100	6,00	6,00	4,00	4,00
0,074	No.200	4,00	4,00	2,00	2,00
	FONDO	0,00	0,00	4,00	4,00
			TOTAL	100,00	100,00

Límite líquido LL	27,00	%
Límite plastico LP	21,00	%
Índice plasticidad IP	6,00	%

Pasa tamiz N° 4 (4,76mm):	70,00	%
Pasa tamiz N° 200 (0,074 mm):	4,00	%
D_{60}:	3,45	mm
D_{30}:	0,62	mm
D_{10} (diámetro efectivo):	0,25	mm
Coeficiente de uniformidad (Cu):	13,78	
Grado de curvatura (Cc):	0,44	

Table 3. Granulometric Composition Thick Floor Example 2.

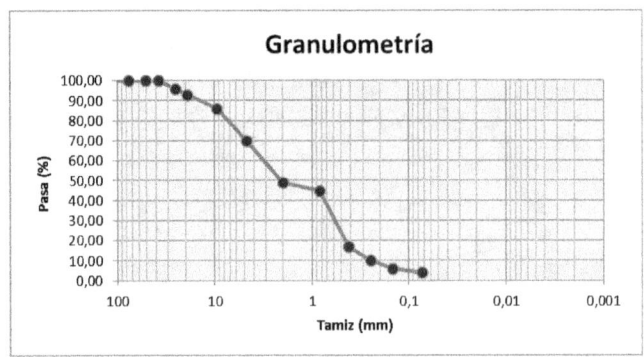

Figure 53. Granulometric Curve Thick Floor Example 2.

Taking into account the results of the atterberg boundaries it can be observed that the indicator point in the plasticity chart

"Carta de Casagrande" is above line A and on the left side of line B, and in the CL-ML area therefore this indicates that the soil is in the area of clays and alms of low plasticity CL-ML , below is the graph of the plasticity chart.

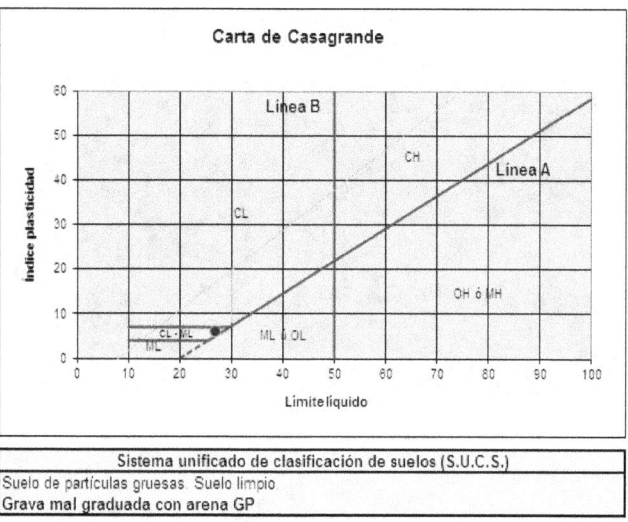

Figure 54. Letter from Casagrande, "Letter of Plasticity" Thick Floor Example 2.

In accordance with the results obtained it can be determined that through the USCS Unified Soil Classification System, the soil belongs to the Coarse **Grain**Soils, Gravas, Clean Gravas, Poorly Graduated Gravas, Gravel-Sand Mixtures, Few Fine or Without Fine and its symbol is **GP,**below is the graph andtable of the USCS Unified Soil Classification System.

CLASIFICACION DE SUELOS USCS

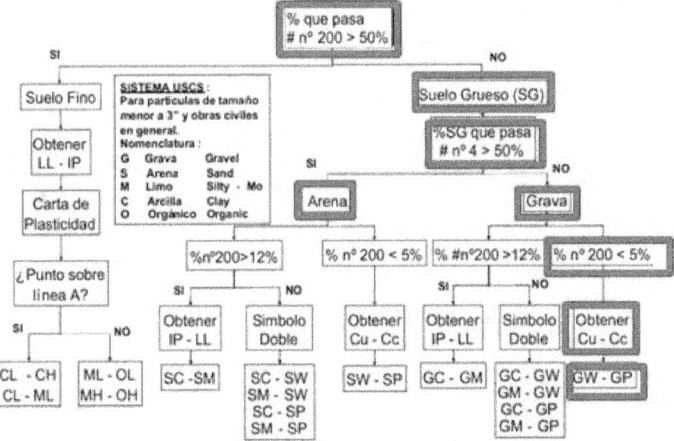

Figure 55. Soil Classification USCS Thick Floor Example 2.

SISTEMA DE CLASIFICACIÓN DE SUELOS UNIFICADO "U.S.C.S."

DIVISIONES PRINCIPALES			Símbolos del grupo	NOMBRES TÍPICOS	IDENTIFICACIÓN DE LABORATORIO		
SUELOS DE GRANO GRUESO Más de la mitad del material retenido en el tamiz número 200	GRAVAS Más de la mitad de la fracción gruesa es retenida por el tamiz número 4 (4.76 mm)	Gravas limpias (sin o con pocos finos)	GW	Gravas, bien graduadas, mezclas grava-arena, pocos finos o sin finos.	Determinar porcentaje de grava y arena en la curva granulométrica. Según el porcentaje de finos (fracción inferior al tamiz número 200). Los suelos de grano grueso se clasifican como sigue:	$Cu=D_{60}/D_{10}>4$ $Cc=(D30)^2/D_{10}xD_{60}$ entre 1 y 3	
			GP	Gravas mal graduadas, mezclas grava-arena, pocos finos o sin finos.		No cumplen con las especificaciones de granulometría para GW.	
		Gravas con finos (apreciable cantidad de finos)	GM	Gravas limosas, mezclas grava-arena-limo.		Límites de Atterberg debajo de la línea A o IP <4.	Encima de línea A con IP entre 4 y 7 son casos límite que requieren doble símbolo.
			GC	Gravas arcillosas, mezclas grava-arena-arcilla.		Límites de Atterberg sobre la línea A con IP>7.	
	ARENAS Más de la mitad de la fracción gruesa pasa por el tamiz número 4 (4,76 mm)	Arenas limpias (pocos o sin finos)	SW	Arenas bien graduadas, arenas con grava, pocos finos o sin finos.	<5%->GW,GP,SW,SP. >12%->GM,GC,SM,SC. 5 al 12%->casos límite que requieren usar doble símbolo.	$Cu=D_{60}/D_{10}>6$ $Cc=(D30)^2/D_{10}xD_{60}$ entre 1 y 3	
			SP	Arenas mal graduadas, arenas con grava, pocos finos o sin finos.		Cuando no se cumplen simultáneamente las condiciones para SW.	
		Arenas con finos (apreciable cantidad de finos)	SM	Arenas limosas, mezclas de arena y limo.		Límites de Atterberg debajo de la línea A o IP <4.	Los límites situados en la zona rayada con IP entre 4 y 7 son casos intermedios que precisan
			SC	Arenas arcillosas, mezclas arena-arcilla.		Límites de Atterberg sobre la línea A con IP>7.	
SUELOS DE GRANO FINO Más de la mitad del material pasa por el tamiz número 200	Limos y arcillas: Límite líquido menor de 50		ML	Limos inorgánicos y arenas muy finas, limos limpios, arenas finas, limosas o arcillosa, o limos arcillosos con ligera plasticidad.	Ábaco de Casagrande		
			CL	Arcillas inorgánicas de plasticidad baja a media, arcillas con grava, arcillas arenosas, arcillas limosas.			
			OL	Limos orgánicos y arcillas orgánicas limosas de baja plasticidad.			
	Limos y arcillas: Límite líquido mayor de 50		MH	Limos inorgánicos, suelos arenosos finos o limosos con mica o diatomeas, limos elásticos.			
			CH	Arcillas inorgánicas de plasticidad alta.			
			OH	Arcillas orgánicas de plasticidad media a elevada; limos orgánicos.			
Suelos muy orgánicos			PT	Turba y otros suelos de alto contenido orgánico.			

Table 4 USCS Unified Soil Classification System Thick Floor Example 2.

The soil is thick because the percentage passed by the sieve No. 200 is less than 50% and belongs to the Group of Sands because the percentage passed by mesh No. 4 is greater than 50%, it is Grava because more than 50% of the thick fraction is retained in the sieve No. 4, is Mal graduated gravel GP because the percentage passing the sieve No. 200 is greater than 5%, through the above information we can obtain the following data.

a. Amount of thick soil: 96%

b. Amount of fine soil: 4%
c. Amount of Gravel: 30%
d. Amount of sand: 66%
e. Amount of coarse gravel: 7%
f. Average amount of sand: 28%
g. Amount of fine sand: 13%
h. Soil classification: GP Poorly graduated gravel, gravel-sand mixtures, few fine or without fine.
i. Plasticity type: CL-ML Clays and low plasticity alms.

Below are 2 examples of fine soil classification using the plasticity chart, by the SUCS system, explain each step.
:

EXAMPLE 1 FINE SOIL

It has a soil to which the corresponding tests have been performed such as granulometry and atterberg limits where the following results were obtained:

Atterberg limits.
LL: Liquid limit: 25%
LP: Plastic limit: 18%
IP: Plasticity Index - LL – LP: 7%

Granulometry.
It ranges from the mesh of sieve No. 4 to mesh No. 200 and has a percentage of material that passes the mesh No. 200 in a range greater than 50% of the total material, below is the table of granulometric composition of the material and graph of the granulometric curve of the soil cited.

Tamiz (mm)	Malla	Pasa (%)	Pasante (%)	Retenido acumulado (%)	Retenido parcial (%)
101,6	4"	100,00	100,00	0,00	0,00
76,2	3"	100,00	100,00	0,00	0,00
50,8	2"	100,00	100,00	0,00	0,00
38,1	1 1/2"	100,00	100,00	0,00	0,00
25,4	1"	100,00	100,00	0,00	0,00
19,1	3/4"	100,00	100,00	0,00	0,00
9,52	3/8"	100,00	100,00	0,00	0,00
4,76	No.4	98,00	98,00	2,00	2,00
2,00	No.10	95,00	95,00	3,00	3,00
0,84	No.20	90,00	90,00	5,00	5,00
0,42	No.40	82,00	82,00	8,00	8,00
0,25	No.60	73,00	73,00	9,00	9,00
0,149	No.100	67,00	67,00	6,00	6,00
0,074	No.200	59,00	59,00	8,00	8,00
	FONDO	0,00	0,00	59,00	59,00
			TOTAL	100,00	100,00

Límite líquido LL	25,00%
Límite plastico LP	18,00%
Indice plasticidad IP	7,00%

Pasa tamiz N° 4 (4,76mm):	98,00 %
Pasa tamiz N° 200 (0,074 mm):	59,00 %
D60:	0,08 mm
D30:	0,04 mm
D10 (diámetro efectivo):	0,01 mm
Coeficiente de uniformidad (Cu):	6,65
Grado de curvatura (Cc):	1,35

Table 5. Granulometric Composition Fine Floor Example 1.

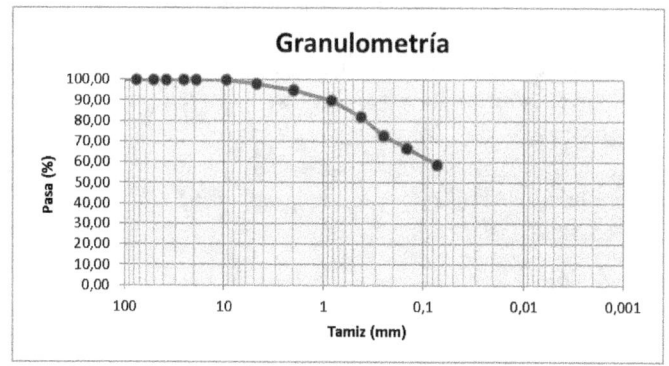

Figure 56. Granulometric Curve Fine Floor Example 1.

Taking into account the results of the atterberg boundaries it can be observed that the indicator point in the plasticity chart

"Carta de Casagrande" is above line A and on the left side of line B, and on the upper line in the CL-ML area therefore this indicates that the soil is in the area of clays and alms of low plasticity CL-ML , below is the graph of the plasticity chart.

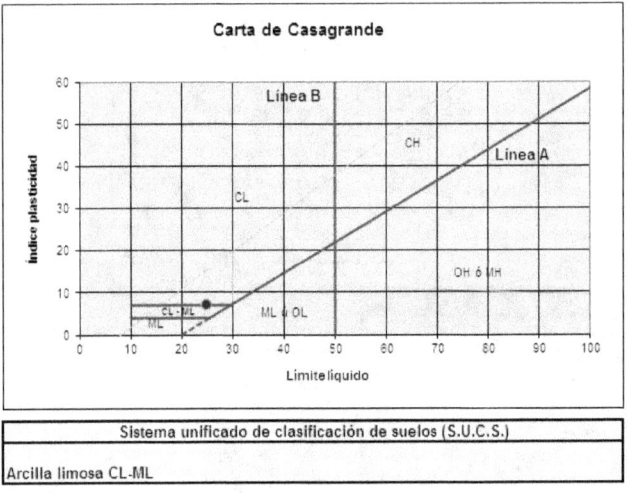

Figure 57. *Letter from Casagrande, "Letter of Plasticity" Fine Floor Example 1.*

In line with the results obtained it can be determined that through the USCS Unified Soil Classification System, the soil belongs to **the Fine Grain**Soils, Limos and Clays, Inorganic clays of low to medium plasticity, clays with gravel, sandy clays, slimy clays, inorganic alms and very fine sands, clean alms, fine sands, slimes or clay, or clay tenders with slight plasticity, and its symbol is double **CL-ML,**below is the graph and table ofthe USCS Unified Soil Classification System.

SOILS AND FOUNDATIONS

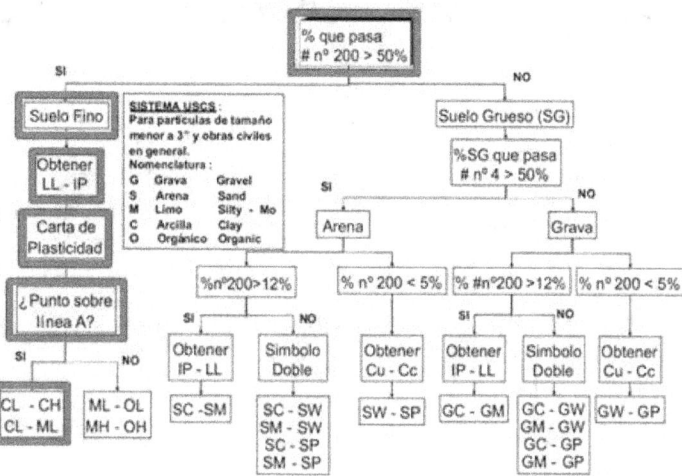

Figure 58. USCS Soil Classification Fine Floor Example 1.

SISTEMA DE CLASIFICACIÓN DE SUELOS UNIFICADO "U.S.C.S."

DIVISIONES PRINCIPALES			Símbolos del grupo	NOMBRES TÍPICOS	IDENTIFICACIÓN DE LABORATORIO		
SUELOS DE GRANO GRUESO Más de la mitad del material retenido en el tamiz número 200	GRAVAS Más de la mitad de la fracción gruesa es retenida por el tamiz número 4 (4,76 mm)	Gravas limpias (sin o con pocos finos)	GW	Gravas, bien graduadas, mezclas grava-arena, pocos finos o sin finos.	Determinar porcentaje de grava y arena en la curva granulométrica. Según el porcentaje de finos (fracción inferior al tamiz número 200). Los suelos de grano grueso se clasifican como sigue:	$C_u = D_{60}/D_{10} > 4$ $C_c = (D_{30})^2/D_{10} \times D_{60}$ entre 1 y 3	
			GP	Gravas mal graduadas, mezclas grava-arena, pocos finos o sin finos.		No cumplen con las especificaciones de granulometría para GW.	
		Gravas con finos (apreciable cantidad de finos)	GM	Gravas limosas, mezclas grava-arena-limo.		Límites de Atterberg debajo de la línea A o IP<4.	Encima de línea A con IP entre 4 y 7 son casos límite que requieren doble símbolo.
			GC	Gravas arcillosas, mezclas grava-arena-arcilla.		Límites de Atterberg sobre la línea A con IP>7.	
	ARENAS Más de la mitad de la fracción gruesa pasa por el tamiz número 4 (4,76 mm)	Arenas limpias (pocos o sin finos)	SW	Arenas bien graduadas, arenas con grava, pocos finos o sin finos.	<5% -> GW,GP,SW,SP. >12% -> GM,GC,SM,SC. 5 al 12% -> casos límite que requieren usar doble símbolo.	$C_u = D_{60}/D_{10} > 6$ $C_c = (D_{30})^2/D_{10} \times D_{60}$ entre 1 y 3	
			SP	Arenas mal graduadas, arenas con grava, pocos finos o sin finos.		Cuando no se cumplen simultáneamente las condiciones para SW.	
		Arenas con finos (apreciable cantidad de finos)	SM	Arenas limosas, mezclas de arena y limo.		Límites de Atterberg debajo de la línea A o IP<4.	Los límites situados en la zona rayada con IP entre 4 y 7 son casos intermedios que precisan
			SC	Arenas arcillosas,		Límites de Atterberg sobre la línea A con IP>7.	
SUELOS DE GRANO FINO Más de la mitad del material pasa por el tamiz número 200	Limos y arcillas: Límite líquido menor de 50		ML	Limos inorgánicos y arenas muy finas, limos limpios, arenas finas, limosas o arcillosa, o limos arcillosos con ligera plasticidad.	Ábaco de Casagrande		
			CL	Arcillas inorgánicas de plasticidad baja a media, arcillas con grava, arcillas arenosas, arcillas limosas.			
			OL	Limos orgánicos y arcillas orgánicas limosas de baja plasticidad.			
	Limos y arcillas: Límite líquido mayor de 50		MH	Limos inorgánicos, suelos arenosos finos o limosos con mica o diatomeas, limos elásticos.			
			CH	Arcillas inorgánicas de plasticidad alta.			
			OH	Arcillas orgánicas de plasticidad media a elevada; limos orgánicos.			
Suelos muy orgánicos			PT	Turba y otros suelos de alto contenido orgánico.			

Table 6 USCS Unified Soil Classification System Fine Soil Example 1.

The soil is thin as the percentage passed by the sieve No. 200 is greater than 50% and belongs to the Group of Sands because the percentage passed by the mesh No. 4 is greater than 50%, it is not Grava because more than 50% of the thick fraction passes the sieve No.4, it is Clay and Slime of low plasticity CL-ML because the percentage passed by the sieve No. 200 is greater than 50%, in addition, the liquid limit is less than 50% using the above information we can obtain the following data.

 a. Amount of thick soil: 41%

b. Amount of fine soil: 59%
c. Amount of Gravel: 2%
d. Amount of sand: 39%
e. Amount of coarse gravel: 0%
f. Average amount of sand: 8%
g. Amount of fine sand: 23%
h. Soil classification: CL-ML Cl-ML Slimy Clay
i. Plasticity type: CL-ML Clays and low plasticity alms.

EXAMPLE 2 FINE SOIL

It has a soil to which the corresponding tests have been performed such as granulometry and atterberg limits where the following results were obtained:

Atterberg limits.
LL: Liquid limit: 58%
LP: Plastic limit: 23%
IP: Plasticity Index - LL – LP: 35%

Granulometry.
It ranges from the mesh of sieve No. 4 to mesh No. 200 and has a percentage of material that passes the mesh No. 200 in a range greater than 50% of the total material, below is the table of granulometric composition of the material and graph of the granulometric curve of the soil cited.

Tamiz (mm)	Malla	Pasa (%)	Pasante (%)	Retenido acumulado (%)	Retenido parcial (%)
101,6	4"	100,00	100,00	0,00	0,00
76,2	3"	100,00	100,00	0,00	0,00
50,8	2"	100,00	100,00	0,00	0,00
38,1	1 1/2"	100,00	100,00	0,00	0,00
25,4	1"	100,00	100,00	0,00	0,00
19,1	3/4"	100,00	100,00	0,00	0,00
9,52	3/8"	100,00	100,00	0,00	0,00
4,76	No.4	98,00	98,00	2,00	2,00
2,00	No.10	90,00	90,00	8,00	8,00
0,84	No.20	83,00	83,00	7,00	7,00
0,42	No.40	78,00	78,00	5,00	5,00
0,25	No.60	70,00	70,00	8,00	8,00
0,149	No.100	67,00	67,00	3,00	3,00
0,074	No.200	65,00	65,00	2,00	2,00
	FONDO	0,00	0,00	65,00	65,00
	TOTAL			100,00	100,00

Límite líquido LL	58,00	%
Límite plastico LP	23,00	%
Índice plasticidad IP	35,00	%

Pasa tamiz N° 4 (4,76mm):	98,00	%
Pasa tamiz N° 200 (0,074 mm):	65,00	%
D_{60}:	0,07	mm
D_{30}:	0,03	mm
D_{10} (diámetro efectivo):	0,01	mm
Coeficiente de uniformidad (Cu):	6,00	
Grado de curvatura (Cc):	1,50	

Table 7 Granulometric Composition Fine Floor Example 2.

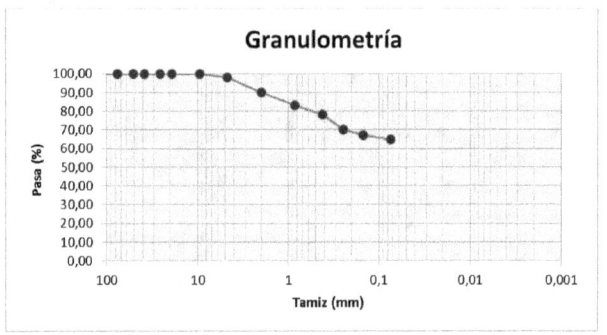

Figure 59. Granulometric Curve Fine Floor Example 2.

Taking into account the results of the atterberg boundaries it can be observed that the indicator point in the plasticity chart

"Carta de Casagrande" is above line A and on the right side of line B, therefore this indicates that the soil is in the area of clays of high plasticity CH, below is the graph of the plasticity chart.

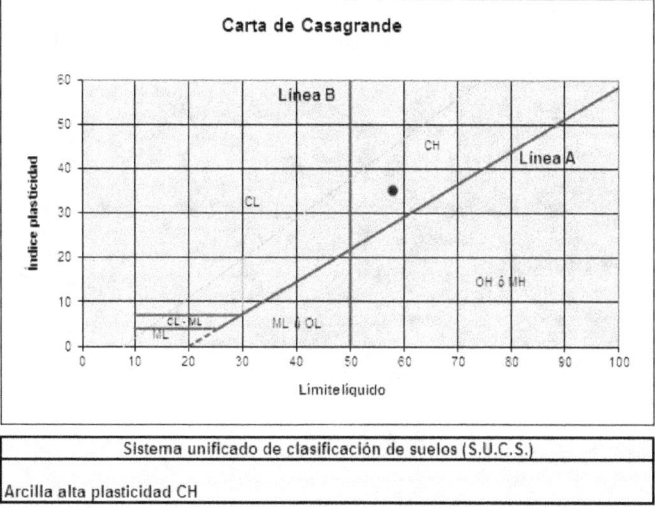

Figure 60. Letter from Casagrande, "Letter of Plasticity" Fine Floor Example 2.

In accordance with the results obtained it can be determined that through the USCS Unified Soil Classification System, the soil belongs to the Fine GrainSoils, Alms and Clays, Inorganic Clays of High Plasticity., and its **symbol is double CH,** below is the graph and table of theUSCS Unified Soil Classification System.

CLASIFICACION DE SUELOS USCS

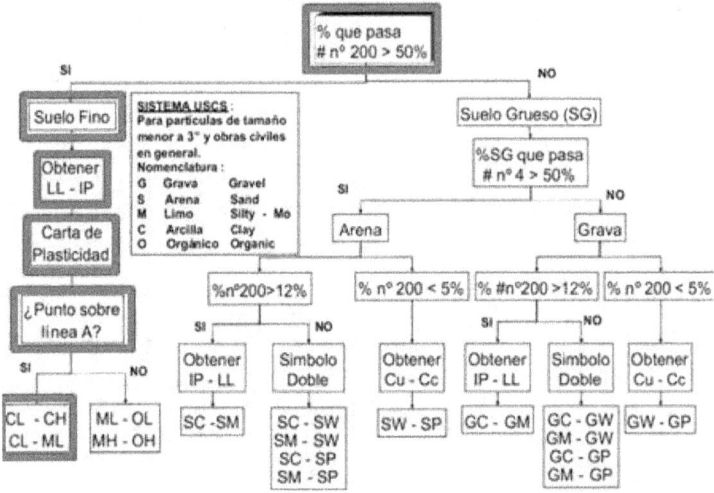

Figure 61. USCS Soil Classification Fine Floor Example 2.

SISTEMA DE CLASIFICACIÓN DE SUELOS UNIFICADO "U.S.C.S."

DIVISIONES PRINCIPALES		Símbolos del grupo	NOMBRES TÍPICOS	IDENTIFICACIÓN DE LABORATORIO		
SUELOS DE GRANO GRUESO — Más de la mitad del material retenido en el tamiz número 200	GRAVAS — Más de la mitad de la fracción gruesa es retenida por el tamiz número 4 (4,76 mm)	Gravas limpias (sin o con pocos finos)	GW	Gravas, bien graduadas, mezclas grava-arena, pocos finos o sin finos.	Determinar porcentaje de grava y arena en la curva granulométrica. Según el porcentaje de finos (fracción inferior al tamiz número 200). Los suelos de grano grueso se clasifican como sigue: <5%->GW,GP,SW,SP. >12%->GM,GC,SM,SC. 5 al 12%->casos límite que requieren usar doble símbolo.	$Cu=D_{60}/D_{10}>4$ $Cc=(D_{30})^2/D_{10}xD_{60}$ entre 1 y 3	
			GP	Gravas mal graduadas, mezclas grava-arena, pocos finos o sin finos.		No cumplen con las especificaciones de granulometría para GW.	
		Gravas con finos (apreciable cantidad de finos)	GM	Gravas limosas, mezclas grava-arena-limo.		Límites de Atterberg debajo de la línea A o IP<4.	Encima de línea A con IP entre 4 y 7 son casos límite que requieren doble símbolo.
			GC	Gravas arcillosas, mezclas grava-arena-arcilla.		Límites de Atterberg sobre la línea A con IP>7.	
	ARENAS — Más de la mitad de la fracción gruesa pasa por el tamiz número 4 (4,76 mm)	Arenas limpias (pocos o sin finos)	SW	Arenas bien graduadas, arenas con grava, pocos finos o sin finos.		$Cu=D_{60}/D_{10}>6$ $Cc=(D_{30})^2/D_{10}xD_{60}$ entre 1 y 3	
			SP	Arenas mal graduadas, arenas con grava, pocos finos o sin finos.		Cuando no se cumplen simultáneamente las condiciones para SW.	
		Arenas con finos (apreciable cantidad de finos)	SM	Arenas limosas, mezclas de arena y limo.		Límites de Atterberg debajo de la línea A o IP<4.	Los límites situados en la zona rayada con IP entre 4 y 7 son casos intermedios que precisan doble símbolo.
			SC	Arenas arcillosas, mezclas arena-arcilla.		Límites de Atterberg sobre la línea A con IP>7.	
SUELOS DE GRANO FINO — Más de la mitad del material pasa por el tamiz número 200	Limos y arcillas: límite líquido menor de 50		ML	Limos inorgánicos y arenas muy finas, limos limpios, arenas finas, limosas o arcillosas, o limos arcillosos con ligera plasticidad.	Ábaco de Casagrande		
			CL	Arcillas inorgánicas de plasticidad baja a media, arcillas con grava, arcillas arenosas, arcillas limosas.			
			OL	Limos orgánicos y arcillas orgánicas limosas de baja plasticidad.			
	Limos y arcillas: límite líquido mayor de 50		MH	Limos inorgánicos, suelos arenosos finos o limosos con mica o diatomeas, limos elásticos.			
			CH	Arcillas inorgánicas de plasticidad alta.			
			OH	Arcillas orgánicas de plasticidad media a elevada; limos orgánicos.			
Suelos muy orgánicos			PT	Turba y otros suelos de alto contenido orgánico.			

Table 8 USCS Unified Soil Classification System Fine Soil Example 2.

The soil is thin as the percentage passed by the sieve No. 200 is greater than 50% and belongs to the Group of Sands because the percentage passed by the mesh No. 4 is greater than 50%, it is not Grava because more than 50% of the thick fraction passes the sieve No.4, it is Clay of high plasticity CH because the percentage passed by the sieve No. 200 is greater than 50% and the liquid limit is greater than 50% using the above information we can obtain the following data.

    a.   Amount of thick soil: 35%
    b.   Amount of fine soil: 65%

c. Amount of Gravel: 2%
d. Amount of sand: 33%
e. Amount of coarse gravel: 0%
f. Average amount of sand: 5%
g. Amount of fine sand: 13%
h. Soil classification: CH Clay of high plasticity
i. Plasticity type: CH Clay high plasticity

CHAPTER 4:

DISTRIBUTION OF STRESSES OR PRESSURES ON THE GROUND UNDER A RIGID SHOE

Efforts and pressures on the ground under a rigid shoe on which a column rests can be determined by the methods listed below.

Boussinesq Theory

The simplest case of the pressure distribution corresponding to a concentrated load, vertically, on the surface of the semi-space, is indicated in the following figure.

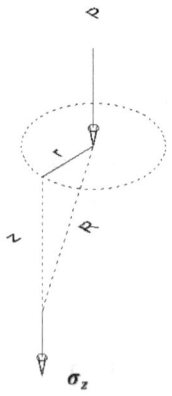

Figure 62. Theoria de Bousinesq.

In the case of soils, the Boussinesq expression that is most interesting is the one that givesthe vertical pressure σ_z, ona horizontal plane at the z-depth and at a radial distance r, that is, the first of the equations.

The most common form of the aforementioned equation is:

$$\sigma_z = \frac{3P}{2 \cdot \pi \cdot z^2} \cdot \frac{1}{\left[1 + \left(\frac{r}{z}\right)^2\right]^{\frac{5}{2}}}$$

Or as follows:

$$\sigma_z = \frac{3P}{2 \cdot \pi \cdot z^2} \cdot \left[\frac{1}{1 + \left(\frac{r}{z}\right)^2}\right]^{\frac{5}{2}}$$

In practice, deformations are studied in the laboratory, extracting unchanged samples from the soil, under the action of efforts.

Through Boussinesq's theory, such efforts can be plott differently. One keeping constant the σ z-shape with which isobaras or pressure bulb is formed.

SOILS AND FOUNDATIONS

DISTRIBUCIÓN DE PRESIONES

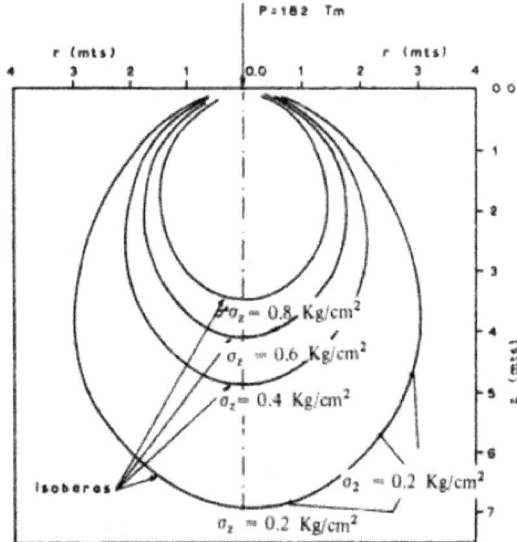

Figure 63 Point Load Pressure Distribution, "Pressure Bulbs or Isobaras"

Another way to plot efforts is by distributing stresses on a horizontal plane to a constant depth "z"

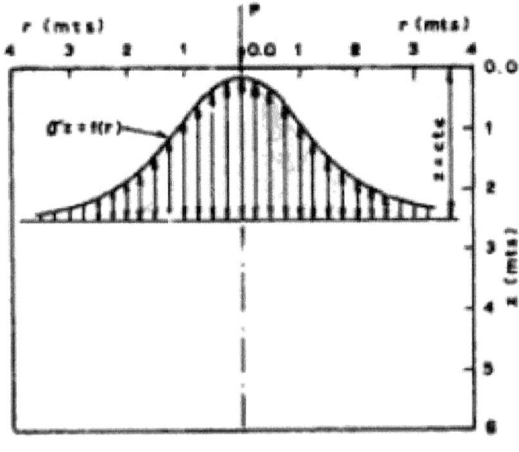

Figure 64. Distribution of Efforts in Horizontal Plane.

Efforts can also be plotted by distributing vertical stresses with depth on a vertical plane at a constant "r" distance from the concentrated vertical load line of action.

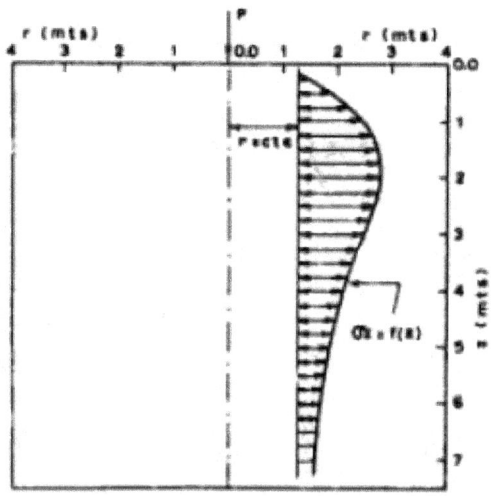

Figure 65. Distribution of Efforts in Vertical Plane.

FADUM table

In 1939, FADUM prepared a table that simplifies the problem based on the author of the integration of the BOUSSINESQ equation for a rectangular surface leaving the point under investigation at a depth *"z"* is given by the equation:

$$\sigma_Z = I \cdot q$$

I - influence value that depend on "m" and *"n"*, taken from the FADUM table

SOILS AND FOUNDATIONS

m- relationship between the width *of* the rectangle and the depth "z"

$$m = \frac{B}{z}$$

n- relationship between the length of the rectangle and the depth "z"

$$n = \frac{L}{z}$$

Tabla 19.2 Valores de *I* para los esfuerzos verticales debajo de una esquina según Fadum

m \ n	0.1	0.2	0.3	0.4	0.5	0.6
0.1	0.00470	0.00917	0.01323	0.01678	0.01978	0.02223
0.2	0.00917	0.01790	0.02585	0.03280	0.03866	0.04348
0.3	0.01323	0.02585	0.03735	0.04742	0.05593	0.06294
0.4	0.01678	0.03280	0.04742	0.06024	0.07111	0.08009
0.5	0.01978	0.03866	0.05593	0.07111	0.08403	0.09473
0.6	0.02223	0.04348	0.06294	0.08009	0.09473	0.10688
0.7	0.02420	0.04735	0.06558	0.08734	0.10340	0.11679
0.8	0.02576	0.05042	0.07308	0.09314	0.11035	0.12474
0.9	0.02698	0.05283	0.07661	0.09770	0.11584	0.13105
1.0	0.02794	0.05471	0.07938	0.10129	0.12018	0.13605
1.2	0.02926	0.05733	0.08323	0.10631	0.12626	0.14309
1.4	0.03007	0.05894	0.08561	0.10941	0.13003	0.14749
1.6	0.03058	0.05994	0.08709	0.11135	0.13241	0.15028
1.8	0.03090	0.06058	0.08804	0.11260	0.13395	0.15207
2.0	0.03111	0.06100	0.08867	0.11342	0.13496	0.15326
2.5	0.03138	0.06155	0.08948	0.11450	0.13628	0.15483
3.0	0.03150	0.06178	0.08982	0.11495	0.13684	0.15550
4.0	0.03158	0.06194	0.09007	0.11527	0.13724	0.15598
5.0	0.03160	0.06199	0.09014	0.11537	0.13737	0.15612
6.0	0.03161	0.06201	0.09017	0.11541	0.13741	0.15617
8.0	0.03162	0.06202	0.09018	0.11543	0.13744	0.15621
10.0	0.03162	0.06202	0.09019	0.11544	0.13745	0.15622
α	0.03162	0.06202	0.09019	0.11544	0.13745	0.15623

Table 9. FADUM table.

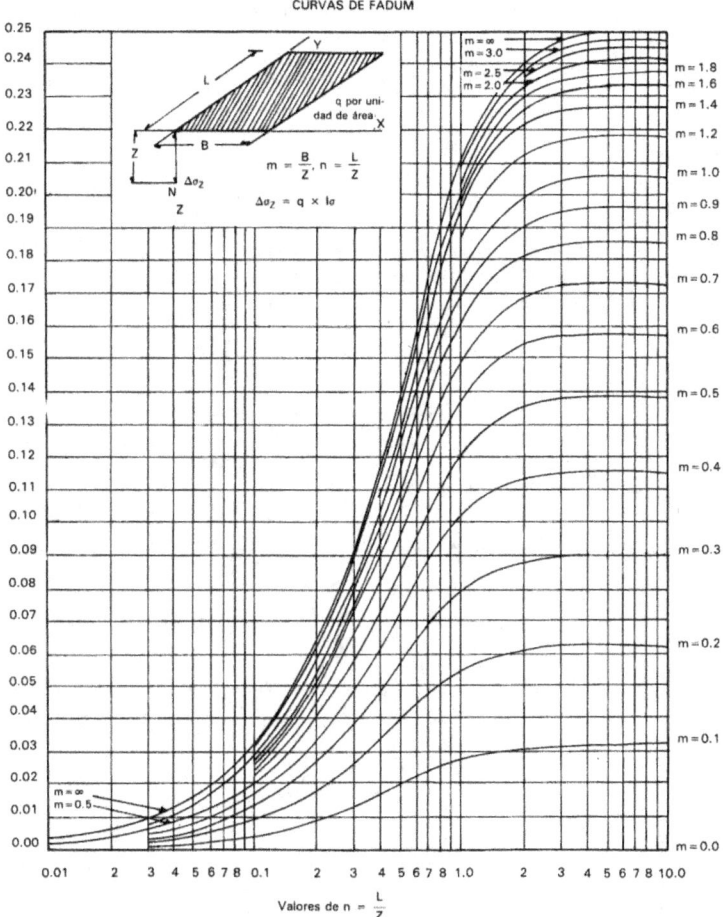

Figure 66. Graph To Determine FADUM Curves.

The vertical pressure under a uniform load on a circular area can be determined directly using table 19.3, **z** and **d** represent, the depth and radial horizontal distance from the center of the circle to the point where the pressure is desired. In **addition R** represents the radius of the circle on which the load acts uniformly, see figure. To calculate vertical pressure, the influence coefficient I is **obtained using** the **z/R and d/R** ratios, and multiplied by the q pressure **applied** to the circular surface

$$\sigma_Z = I \cdot q$$

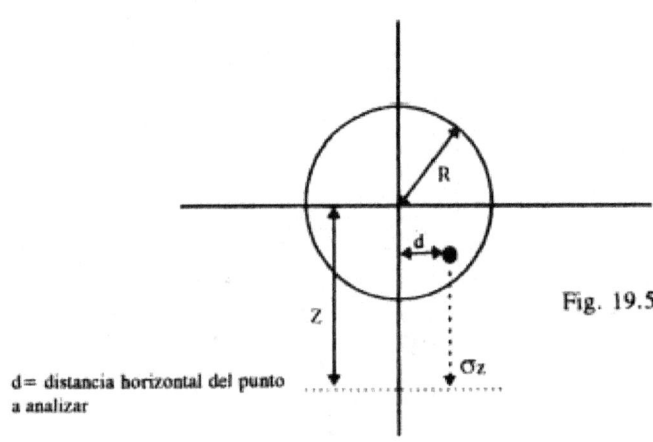

Fig. 19.5

d= distancia horizontal del punto a analizar

Figure 67. Vertical Pressure Under a Uniform Load On a Circular Area.

The **I** is obtained from Table 19.3

Tabla 19.3

Z/R	d/R									
	0	0.3	0.5	1.0	1.5	2.0	2.5	3.0	3.5	4.0
0.25	0.986	0.983	0.964	0.460	0.015	0.000	0.002	0.000	0.000	0.000
0.50	0.911	0.895	0.840	0.418	0.060	0.010	0.003	0.000	0.000	0.000
0.75	0.784	0.762	0.691	0.374	0.105	0.025	0.010	0.002	0.000	0.000
1.00	0.646	0.625	0.560	0.335	0.125	0.043	0.016	0.007	0.003	0.000
1.25	0.524	0.508	0.455	0.295	0.135	0.057	0.023	0.010	0.005	0.001
1.50	0.424	0.413	0.374	0.256	0.137	0.064	0.029	0.013	0.007	0.002
1.75	0.346	0.336	0.309	0.223	0.135	0.071	0.037	0.018	0.009	0.004
2.00	0.284	0.277	0.258	0.194	0.127	0.073	0.041	0.022	0.012	0.006
2.50	0.200	0.196	0.186	0.150	0.109	0.073	0.044	0.028	0.017	0.011
3.00	0.146	0.143	0.137	0.117	0.091	0.066	0.045	0.031	0.022	0.015
4.00	0.087	0.086	0.083	0.076	0.061	0.052	0.041	0.031	0.024	0.018
5.00	0.057	0.057	0.056	0.052	0.045	0.039	0.033	0.027	0.022	0.018
7.00	0.030	0.030	0.029	0.028	0.026	0.024	0.021	0.019	0.016	0.015
10.00	0.015	0.015	0.014	0.014	0.013	0.013	0.013	0.012	0.012	0.110

Table 10. Table To Determine The Influence Coefficient "I".

Newmark method

Another way to determine vertical stresses, produced at a certain depth, due to surface loads, is to make use of **the N.M. Newmark.**

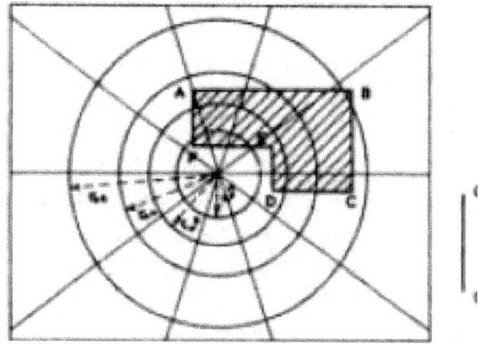

Figure 68. Newmark Method Pressure Distribution.

The vertical stress at a z-depth under the center of a uniformly charged circular radius r area is: σ_z

$$\sigma_z = I \cdot q = \left(1 - \left(\frac{1}{1 + \left(\frac{r}{z}\right)^2}\right)^{\frac{3}{2}}\right) q$$

In which q *is* the unit load on the circle and the value of I is:

$$I = 1 - \left(\frac{1}{1 + \left(\frac{r}{z}\right)^2}\right)^{\frac{3}{2}}$$

From the ec. Previous that gives the vertical stress value to a given depth, you can determine the value of $(\sigma_z r/z)$ that corresponds to because; $\frac{\sigma_z}{q} = 0.8$

$$\frac{\sigma_z}{q} = 1 - \left(\frac{1}{1 + \left(\frac{r}{z}\right)^2}\right)^{\frac{3}{2}}$$

And it turns out that (r/z) is equal to *1,387*

The procedure for using the *Newmark diagram* is:
The plane of the foundation is drawn where the **OQ segment** **of** the abac represents the z depth **of** the point at which the effort is to be known. The abac is placed on the foundation drawing so that the projection of the point being studied coincides with the **$\sigma_z O$ center of** the abac, is the number of areas covered by the foundation area and the product of this number by the coefficient of influence of each zone and by the value of q **provides** the value of at the point considered σ_z

SOILS AND FOUNDATIONS

In all cases, the procedure to be followed has to be defined by the engineer who designs since the type of work and the type of project will be aspects that you have to take into account to choose the procedure that you think is most appropriate

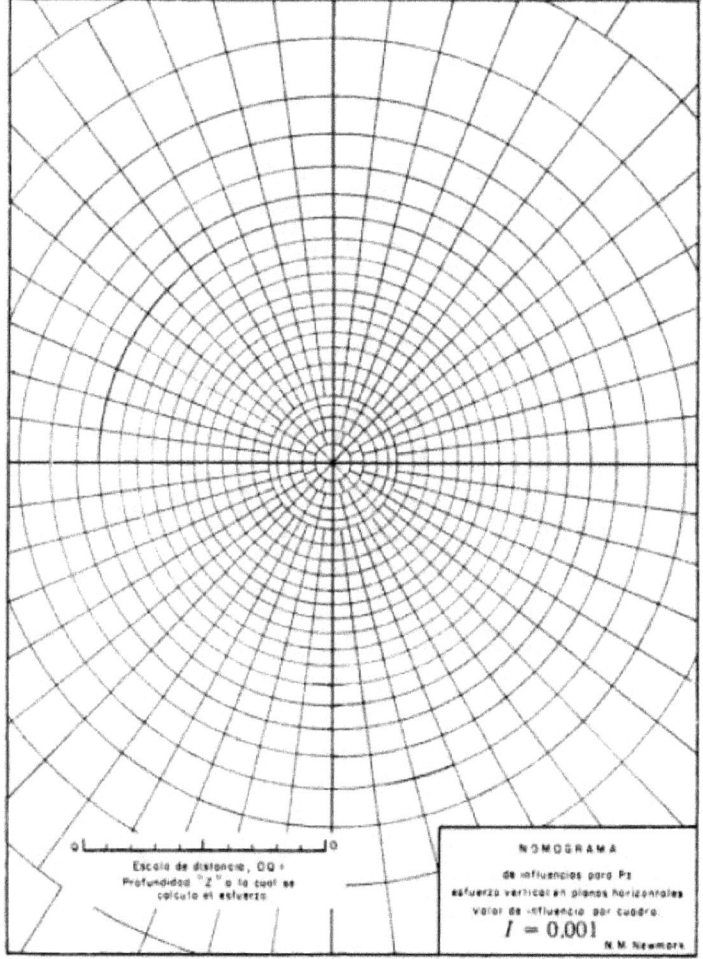

Figure 69. Newmark Nomogram.

Method two in one

In many cases a simple method can be followed to determine the approximate pressure, a method called 2 in 1, in which the load is assumed to be distributed under a slope of twice the height at once the base. If we assume that at ground level a structure has dimensions $\sigma_z A$ *and* B, at a depth z, the weight of the structure will be divided over an area of sides A + z *and* B + z. the maximum pressure is estimated 1.5 times the previous one which is the mean.

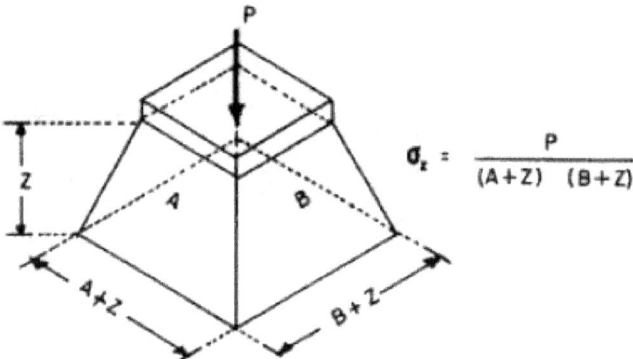

Figure 70. *Two-in-One Method Pressure Distribution.*

Example.
Calculate the pressure at a point 5 m deep below the center of a foundation 6 m x 20 m long that supports a uniform load of 2 kg/cm²

Solution.
Average Pressure 0.87 Kg/Cm², Maximum Pressure - 1.31 Kg/Cm²

Total load: 20 Ton/m² x 6 m x 20 m x 2400 tons.
The area of distribution of this load at a depth of 5 m is:
Distribution area (6+5)(20+5) s 11x25 x 275 m²
Thus the average pressure (not the maximum) at that depth shall be:

$$\sigma_z = \frac{P}{A}$$

$$\sigma_z = \frac{2400}{275} \cdot 8.7 \text{ Ton/m}^2 \text{ x } 0.87 \text{ kg/cm}^2$$

The estimated maximum pressure shall be:

$$\sigma_z = 0.87 x\, 1.5 = 1.31\, kg/cm2$$

Knowing how pressures are distributed in soils, it is advisable to know the resistance of the different strata in order to define whether or not harmful settlements occur when placing new loads.

DEAD AND LIVE LOAD PER COLUMN

BASED ON THE AFFERENT AREA.

The transmission of loads to the foundation is produced by the relationship of light between supports of the roof slab or mezzanine, and by different conditions that may influence its supports, either by load walls, beams and these in turn on columns or edge elements.

The loads transmitted through the structure are of 3 types which are living loads, dead loads and environmental loads, with living loads being those that influence externally the structure and are produced by the use and occupation of the building, the dead loads that make up the weight of the structure itself and the

SOILS AND FOUNDATIONS

environmental loads are those produced by the action of the natural environment such as wind, earthquake, tsunami, rain, the legal regulations in force in Colombia that govern loads and their combinations and adjustments for structure design is TITLE B of the NSR-10 Colombian Regulations on Sismoresisnt Constructions.

Tabla B.3.2-1
Masas de los materiales

Material	Densidad (kg/m^3)	Material	Densidad (kg/m^3)
Acero	7 800	Mortero de inyección para mampostería	2 250
Agua		Mortero de pega para mampostería	2 100
Dulce	1 000	Piedra	
Marina	1 030	Caliza, mármol, cuarzo	2 700
Aluminio	2 700	Basalto, granito, gneis	2 850
Arena		Arenisca	2 200
Limpia y seca	1 440	Pizarra	2 600
Seca de río	1 700	Plomo	11 400
Baldosa cerámica	2 400	Productos bituminosos	
Bronce	8 850	Asfalto y alquitrán	1 300
Cal		Gasolina	700
Hidratada suelta	500	Grafito	2 160
Hidratada compacta	730	Parafina	900
Carbón, apilado	800	Petróleo	850
Carbón vegetal	200	Relleno de ceniza	920
Cemento pórtland, a granel	1 440	Tableros de madera aglutinada	750
Cobre	9 000	Terracota	
Concreto simple	2 300	Poros saturados	1 950
Concreto reforzado	2 400	Poros no saturados	1 150
Corcho, comprimido	250	Tierra	
Estaño	7 360	Arcilla húmeda	1 750
Grava seca	1 660	Arcilla seca	1 100
Hielo	920	Arcilla y grava seca	1 600
Hierro		Arena y grava húmeda	1 900
Fundido	7 200	Arena y grava seca apisonada	1 750
Forjado	7 700	Arena y grava seca suelta	1 600
Latón	8 430	Limo húmedo consolidado	1 550
Madera laminada	600	Limo húmedo suelto	1 250
Madera seca	450-750	Vidrio	2 600
Mampostería de concreto	2 150	Yeso en tableros para muros	800
Mampostería de ladrillo macizo	1 850	Yeso suelto	1 150
Mampostería de piedra	2 200	Zinc en láminas enrolladas	7 200

Table 11. Mass of Materials.

Tabla B.4.2.1-1
Cargas vivas mínimas uniformemente distribuidas

Ocupación o uso		Carga uniforme (kN/m²) m² de área en planta	Carga uniforme (kgf/m²) m² de área en planta
Reunión	Balcones	5.0	500
	Corredores y escaleras	5.0	500
	Silletería fija (fijada al piso)	3.0	300
	Gimnasios	5.0	500
	Vestíbulos	5.0	500
	Silletería móvil	5.0	500
	Áreas recreativas	5.0	500
	Plataformas	5.0	500
	Escenarios	7.5	750
Oficinas	Corredores y escaleras	3.0	300
	Oficinas	2.0	200
	Restaurantes	5.0	500
Educativos	Salones de clase	2.0	200
	Corredores y escaleras	5.0	500
	Bibliotecas Salones de lectura	2.0	200
	Estanterías	7.0	700
Fábricas	Industrias livianas	5.0	500
	Industrias pesadas	10.0	1000
Institucional	Cuartos de cirugía, laboratorios	4.0	400
	Cuartos privados	2.0	200
	Corredores y escaleras	5.0	500
Comercio	Minorista	5.0	500
	Mayorista	6.0	600
Residencial	Balcones	5.0	500
	Cuartos privados y sus corredores	1.8	180
	Escaleras	3.0	300
Almacenamiento	Liviano	6.0	600
	Pesado	12.0	1200
Garajes	Garajes para automóviles de pasajeros	2.5	250
	Garajes para vehículos de carga de hasta 2.000 kg de capacidad.	5.0	500
Coliseos y Estadios	Graderías	5.0	500
	Escaleras	5.0	500

Table 12. Uniformly Distributed Minimum Living Loads.

INCREASED LOAD COMBINATIONS USING

THE RESISTANCE METHOD

LOAD COMBINATIONS

In accordance with NSR-10 TITLE (B.2.4.2) the design of structures, their components and foundations should be done in such a way that their design resistors equal or exceed the effects produced by the higher loads in the following combinations:
1.4(D+ F) (B.2.4-1)
1.2(D+F+T) + **1.6(L+H)** + **0.5**(Lr**or G or Le)
(B.2.4-2)**
1.2D+ 1.6(Lr or G**or Le**) + (L**or 0.8W) (B.2.4-3)**
1.2D+ 1.6W+ 1.0L + 0.5(Lr or G or**Le**) **(B.2.4-4)**
1.2D+ 1.0E + 1.0L (B.2.4-5)

0.9D+ 1.6W+ 1.6H (B.2.4-6)
0.9D+ 1.0E + 1.6H (B.2.4-7)

Where:
D - Dead load consisting of:
(a) the element's own weight.
(b) weight of all building materials incorporated into the building and which are
permanently supported by the element, including walls and dividing partitions of spaces.
(c) weight of permanent equipment.
E - reduced seismic design forces (**E** s **Fs R**) that are used to designthe members
Structural.
Ed - seismic force of the damage threshold.
F - loads due to the weight and pressure of fluids with well-defined densities and controllable maximum heights.
F - load due to flooding.
Fs – seismic forces calculated in accordance with the requirements of Title A of the Regulations.
G - load due to hail, without regard to the contribution of recessing.
L - live loads due to the use and occupation of the building, including loads due to moving objects, partitions that can be changed. **L** includes any reductions that are allowed. If impact load resistance is taken into account this effect should be taken into account in live load **L** .
The water recessed load.
Lr - live charge on the deck.
L0 - undeded live load, in kN/m2. See B.4.5.1.
H - loads due to lateral thrust of soil, groundwater or materials stored with restriction
Horizontal.
R0 - basic energy dissipation capacity coefficient defined for each structural system and each
degree of energy dissipation capacity of the structural material. See Chapter A.3.
R - coefficient of energy dissipationcapacity to be used in the design, corresponds to the basic energy dissipation coefficient

multiplied by the coefficients of reduction of energy dissipation capacity by irregularities in height and in plant, and by absence of redundancy in the structural system of seismicresistance (**R** s **φtoφpφrR0**). See Chapter A.3.

T - forces and effects caused by cumulative effects of temperature variation, setting retraction,

plastic flow, moisture changes, differential settlement or combination of several of these effects.

W - Wind charge.

The following is an example of transmitting live and dead loads based on the afferent area for a column:

EXAMPLE TRANSMISSION OF LOADS.

For the analysis of the structure mentioned below, we must initially see it as a portico by which we can distribute the loads presented on it.

FLOOR VIEW

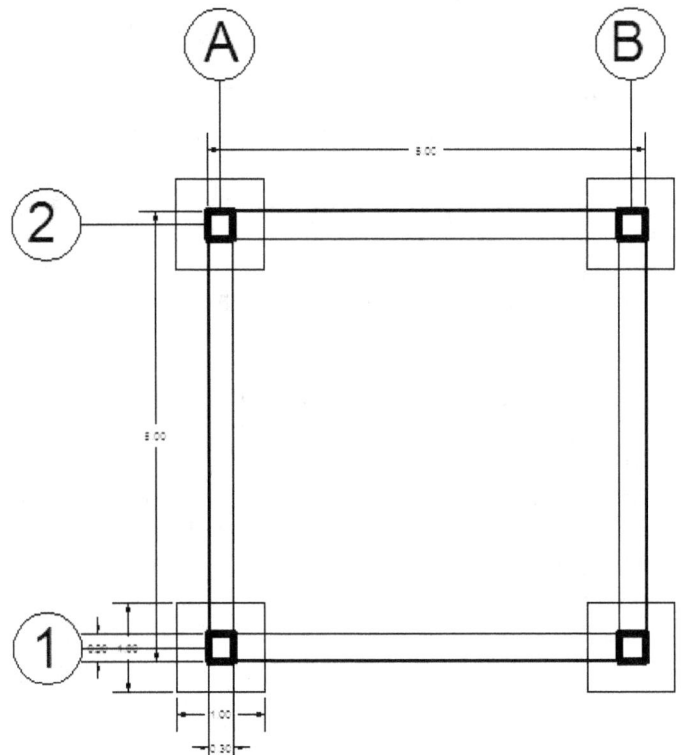

Figure 71. Plan View of Structure to Analyze.

RISING STRUCTURE

SOILS AND FOUNDATIONS

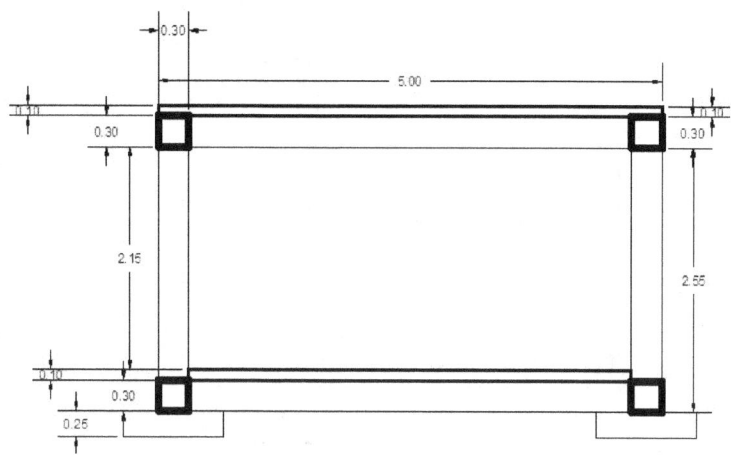

Figure 72. Structure Elevations to Analyze.

BEAM AND COLUMN SECTION

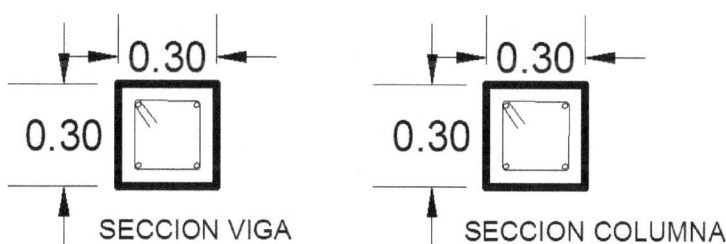

Figure 73. Sections of Beams and Columns Structure to Analyze.

SSA DIMENSIONS

SOILS AND FOUNDATIONS

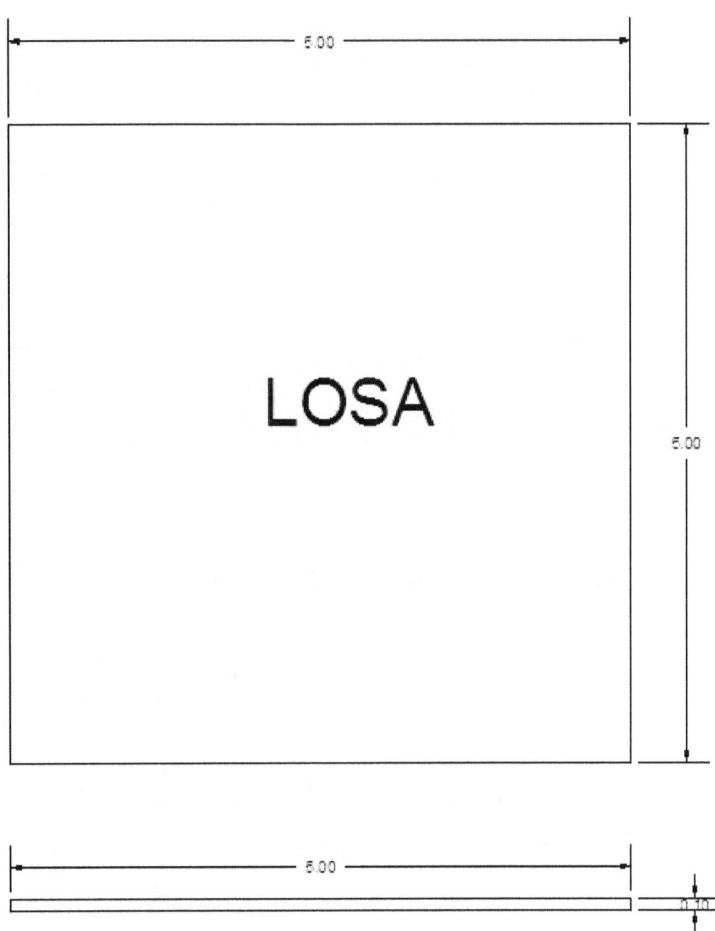

Figure 74. Structure Sn sn sn sn ss to analyze.

DISTRIBUTION OF LOADS BY AFFERENT AREA

SOILS AND FOUNDATIONS

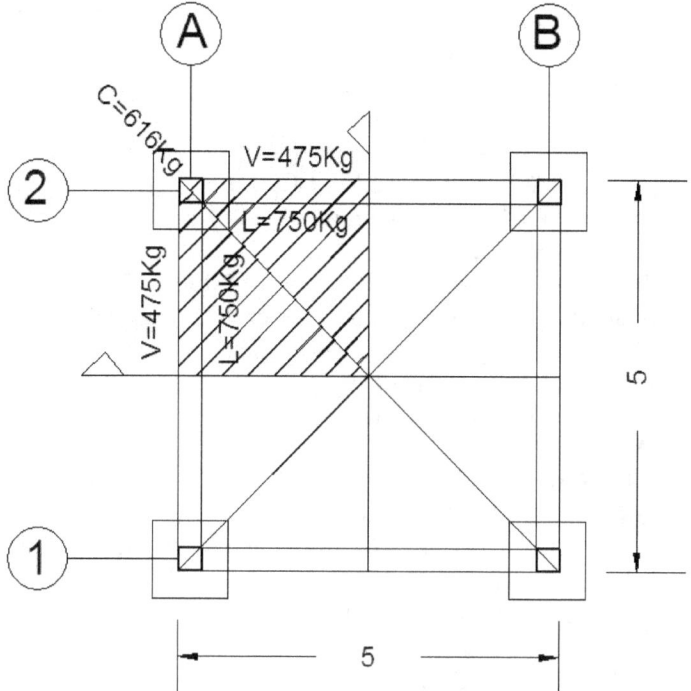

Figure 75. Distribution of Loads by Afferent Area of Structure Column to Be Analyzed.

According to the dimensions mentioned we have that the corresponding weight for each structural element taking into account that they will be reinforced concrete which has a density of 2400 kg/m³ is the one quoted below:

LOSA: 5m*5m*0.10m x 2.5m³ *(2400 Kg/m³) x6000 Kg
VIGAS: 0.30m*0.30m*4.40m x 0.396m³ *(2400 Kg/m³)x950.4 Kg * 4 und x 3802 Kg
COLUMNS: 0.30m*0.30m*2.85m x 0.256m³ *(2400 Kg/m³)x615.6 Kg * 4 und x 2462 Kg

TOTAL WEIGHT: 6000 Kg + 3802 Kg + 2462 Kg x 12264 Kg

WEIGHT THAT LOADS EACH COLUMN PER AFFERENT AREA:

12264 Kg - 2462 Kg x 9802 Kg / 4 und s 2450 Kg

DEAD LOAD "DL":

TOTAL WEIGHT DEAD LOAD: 6000 Kg (sly) + 3802 Kg (beams) + 2462 Kg (columns) - 12264 Kg

LIVE CHARGE "LL":

In co-equity with Table B.4.2.1-1 of NSR-10 TITLE B and assumingthat the area in which the loads are being analyzed is for Residential use we take as living load value 180 $Kgf/m^{2\ measured}$ in plant, so we have that the living load of our analyzed structure is the one obtained from the next operation.

Area on the floor - 5m * 5m x 25 m^2

Live Load "LL" - 25 m^2 * 180 Kgf/m^2 x 4500 Kg

SERVICE CHARGE qs

The service load is obtained from the summation between live loads and dead loads, so we have to.

Service Load "qs" - DL+LL - 12264 Kg + 4500 Kg - 16764 Kg

MAJOR CHARGES

The total load is increased taking into account both live loads, dead loads and environmental loads in accordance with the load combination that applies to the case according to NSR-10 TITLE (B.2.4.2).

In this case we will use the load combination:

1.2(D+F+T)   + **1.6(L+H)**   + **0.5**(L_r or **G** or **L**e)   **(B.2.4-2)**

Where:

D - Dead load consisting of:

(a) the element's own weight.

(b) weight of all building materials incorporated into the building and which are

permanently supported by the element, including walls and dividing partitions of spaces.

(c) weight of permanent equipment.

F - loads due to the weight and pressure of fluids with well-defined densities and controllable maximum heights.

SOILS AND FOUNDATIONS

G - load due to hail, without regard to the contribution of recessing.

L - live loads due to the use and occupation of the building, including loads due to moving objects, partitions that can be changed. **L** includes any reductions that are allowed. If impact load resistance is taken into account this effect should be taken into account in live load **L** .

The water recessed load.

Lr - live charge on the deck.

H - loads due to lateral thrust of the soil, groundwater or horizontally restricted stored materials.

T - forces and effects caused by cumulative effects of temperature variation, setting retraction,

plastic flow, moisture changes, differential settlement or combination of several of these effects.

We replace the data with which they will influence our area to be analyzed as follows:

1.2(D "Dead") + **1.6(L** "Live") + **0.5(Lr**"Live onDeck"))
1.2*(12264 Kg) + **1.6***(4500 Kg)+ **0.5***(0)
14717 Kg + 7200 Kg + 0 Kg x 21917 Kg

Final Load - 21917 Kg

Load per m^2 x 21917 Kg / 25 m^2 x 877 Kg/m^2

Total Bulk load received for each column based on its afferent area per floor:

21917 Kg /4 Columns - **5479 Kg** "includes own weight of column"

54.79 KN

LAST LOAD CAPACITY qu

It is defined as the last pressure per unit of foundation area supported by the ground, in excess of the pressure caused by the soil around the level of the foundation. If the difference between the specific weight of the material that makes up the foundation and the specific weight of the soil surrounding it is supposed to be negligible, then:

q net ? qu – q

Last net load capacity

SOILS AND FOUNDATIONS

In conclusion, it is the maximum contact effort that can withstand a soil mass without a catastrophic settlement of the foundation and the structure supported by it, which corresponds to the theoretical value of the maximum support capacity of a soil. In the calculation of support capacity, account should be taken of soil shear resistance at foundation level and settlements likely by consolidation effect.

$$qu = cNc + \gamma Df$$

Where:
c (ton/m²) - Soil cohesion.
Nc - Load capacity factor.
γ (ton/m³) - Volumetric weight of soil mass.
Df (m) - Depth of shoe flattening.
Fs - Safety factor.

PERMISSIBLE LOAD CAPACITY qa

Vmaximum contact effort applicable to the design and construction of a foundation. The permissible support capacity is only a fraction of the ultimate support capacity, and is calculated by applying an appropriate safety factor to the latter. The most common factor of safety values are in the range of 3 to 5.

$$qa = \frac{cNc}{Fs} + \gamma Df$$

Where:
c (ton/m²) - Soil cohesion.
Nc - Load capacity factor.
γ (ton/m³) - Volumetric weight of soil mass.
Df (m) - Depth of shoe flattening.
Fs - Safety factor.

SIZING AN ISOLATED SHOE AND A RUN.

EXAMPLE ZAPATA RUN FOR WALL.

ELASTIC THEORY

Example:
It is required to calculate a running shoe for a concrete wall

that transmits a uniformly distributed load of 13 tons per meter.
Data:
Width of the wall x 30 cm.
Allowable load of the foundation run σ_{to} 1.0 kg/cm² x 10 Tm/m².
γ cx 2500 kg/m³ x 2.5 Tm/m³
fs x 1400 kg/cm² (fy x 2800 kg/cm²)
f'c s 175 kg/cm²; fc x 0.45 x f'c s 79 kg/cm²
n x 12
Allowable effort at cutting - V_{ad} - 0.29 0.29 x 3.84 kg/cm $\sqrt{f'c} = \sqrt{175} = 2$, when analyzed as a beam.

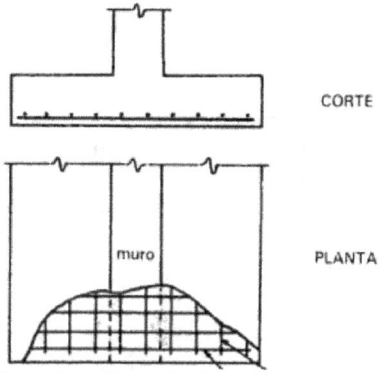

Figure 76. Cut and Plant Zapata Corrida.

CALCULATION OF THE NET REACTION

To calculate the net soil reaction, an effective amount of 15 cm, which plus the 7.0 cm coating gives a total supereal h x 15 + 7.0

22.0 cm. the net reaction will have a value of: σ_n s σ_a - σ_c, s10 - (0.22 x 2.5) - 10 - 0.55 x 9.45 Tm/m² x 0.945 kg/cm²
Considering a unit length of sn sn ss you have:

$\sigma_{n'}$; $= \dfrac{P}{A} = \dfrac{P}{B.L} BL = \dfrac{P}{\sigma n}, y$ como $L = 1,00m$
you have:

$$B = \frac{P}{\sigma n} = \frac{13}{9{,}45} = 1.38 \text{ m. It will be used } 1.40 \text{ m.}$$

When used as a value of B x 1.40 m, instead of 1.38 m, you will have a new
sigma net. Therefore:

$$\sigma n = \frac{13}{1{,}4(1{,}0)} \cdot 9.3 \text{ Tm/m}^2 \text{ x } 0.93 \text{ kg/cm}^2$$

CALCULATION OF THE PROFILE AND THE

FLECTOR MOMENT REINFORCEMENT

"HEIGHT"

It has already been said that the critical section for the flector moment and for adhesion is the same. and for this example, as sf there is dala or contraction under the wall, it is one defined by the vertical plane that passes through the parament of the
wall (see image).

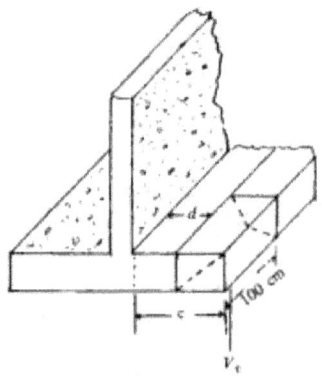

Figure 77. Cumshot Shoe Scheme.

The moment flector considering a meter of wall (L x 1.0 m) will be worth:

SOILS AND FOUNDATIONS

$$M = \sigma n * c * L * \frac{c}{2} = \frac{\sigma n * c^2}{2} = \frac{9.3 * 0.55^2}{2}$$
$$= 1.407 \text{ Tm} - m = 1407 \text{ kg} - m$$
$$= 140{,}700 \text{ kg} - cm$$

For the balanced section the constants are:
k x 0.407; j x 0.864 and K x 14.06 kg/cm2

$$d = \frac{\sqrt{M}}{K * b} = \frac{\sqrt{140.700}}{14.06 * 100} = 10 \text{cm}$$

As the pearl at the moment came out less than 15 cm. which is the minimum advisable, and that has been the assumption, it will work with 15 cm.

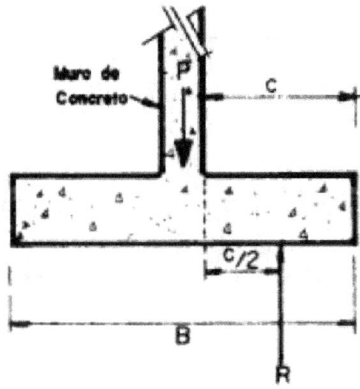

Figure 78. Cumshot Cumshot.

EXAMPLE OF ISOLATED SHOE.

It is intended to design an insulated shoe with the following information:

Service load P x 344 kN

Allowable load qa x 100 kN/m2
F'c' 21MPa
Fy x 420 MPa

b1 x 300 mm
b2 x 400 mm

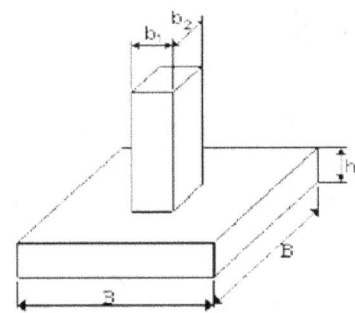

Figure 79. Insulated shoe.

The foundation elements are sized to withstand the higher loads and induced reactions. The support area of the foundation's base is determined by unsteer forces and permissible effort on the ground.

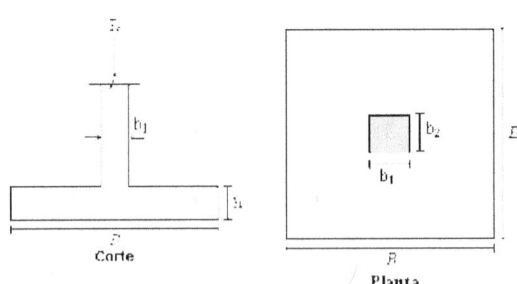

Figure 80. Cut and Insulated Shoe Plant.

Sizing

The service load is:
Ps x 344kN

The permissible capacity of the soil is:

qa x 100kN/m 2

Therefore B will be given by equation (9):

$$B = \sqrt{\frac{P_s}{q_a}}$$

$$B = \sqrt{\frac{344 \text{kN}}{100 \text{kN/m}^2}}$$

$$B = 1.85 \text{ m}$$

SOILS AND FOUNDATIONS

Punch cutter critical section to "d/2" of the column (bidirectional cutter)

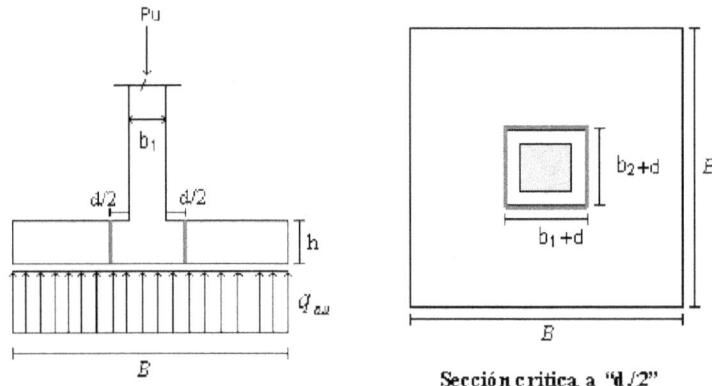

Figure 81. Punch cutter critical section to "d/2" of the column (bidirectional shear.

The thickness of the shoe above the lower reinforcement should not be less than 150 mm for shoes on the ground (C.15.7-NSR-10).
Initially a shoe thickness of:

h x 250 mm

The minimum coating should be 75 mm, in case the shoe is supported on natural soil in accordance with (C.7.7.1-NSR-10)
The effective depth for a 75 mm coating is:

d s h -75 mm

d x 250 mm - 75 mm

d x 175 mm > 150 mm

It is greater than 150 mm so it meets NSR-10 regulatory requirements

(IF CHECK)

As it is a concrete structure, the ultimate load is approximately equal to the service load multiplied by 1.5; This is:

Pu x 1.5 *P x 516 kN

The ultimate effort applied to the foundation floor for the structural design of the shoe is:

$$qu = \frac{Pu}{B^2}$$

$$qu = \frac{516 \text{ KN}}{(1.85m)^2}$$

$$qu = 151 \text{ KN/m}^2$$

Through the results obtained it can be determined that the dimensions of the shoe will be:
B'1 .85 m
L x 1.85 m
0.25 m.

CHAPTER 5: TYPES OF FOUNDATION FAILURES

In line with the experiences and observations regarding the behavior of the foundations it has been seen that the failure due to the load capacity of these happens as a result of a rupture bycutting the floor of the foundation flattening, according to this it can be said that there are three types of failure under the foundations which are cited below.

GENERAL CUT FAILURE

- Well-defined fault pattern (floor wedge and two continuous sliding surfaces within the terrain), the sliding surface starts at the edge of the foundation and advances to the surface of the terrain.
- The surface of the terrain to the shoe rises and can rotate (tilting)

- The fault is violent and catastrophic.
- This fault is typical of dense sand, hard clays, firm cohesive soils and compacted sands when the foundation is deplanted at shallow depths.

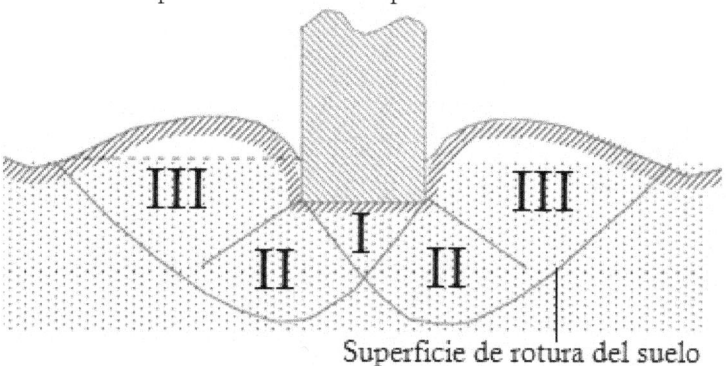

Figure 82. General cut-off fault scheme: I-wedge of elastic state, II - active state zone, III - zone in passive state.

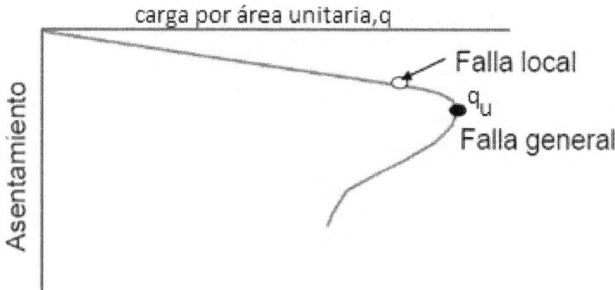

Figure 83. Settlement / Load by unit area q.

When the load per unit area is equal to a sudden ground failure that supports foundation, and the ground fault surface will extend to the ground surface.q_u

LOCAL CUTTING FAILURE

- Fault pattern is only well defined under the shoe.
- A foundation settlement is clearly defined but smaller than in the punch-and-table fault.
- Visible tendency to lift the ground around the shoe.
- There will be no catastrophic collapse of the shoe or a rotation of the shoe.

- This form of breakage is applied if the foundation rests on sand or clay soil of medium compaction
- It is a transitional mode between general failure and punch failure.

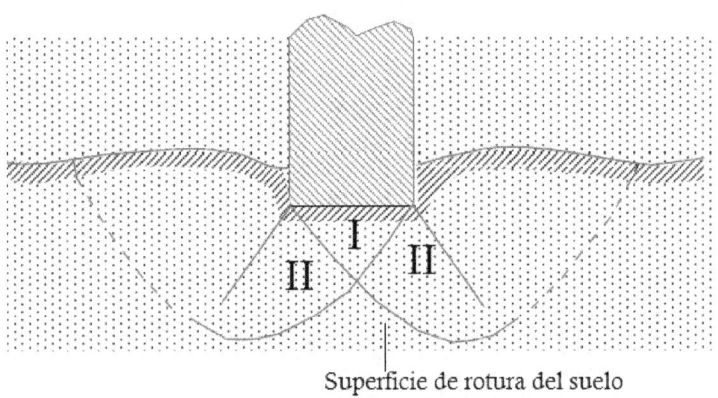

Figure 84. *Local cutting failure scheme: I ' elastic state wedge, II ' active state zone.*

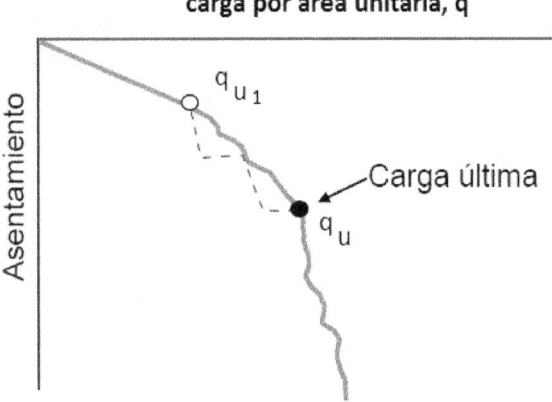

Figure 85. *Settlement / Load by unit area q.*

When the load per unit area on the foundation is equal to qu1 (first failure load), the movement of the foundation will be accompanied by sudden jolts. A considerable foundation movement is then required for the fault surface to extend to the ground surface. The unit area load at which this occurs is the ultimate load capacity (qu) in which a sudden sinking occurs, from which the increased load will result in large seat increments, with no clear sinking load appearing. Also, note that a maximum value of q does not occur in this type of failure.

PUNCH FAILUREMIEMTO

- This fault is characterized by a pattern that is not easily observable when increasing load.
- The vertical movement of the foundation is due to volumetric compression of the ground under it and when the sinking increases a vertical rupture occurs, by cutting around the foundation.
- As the load continues to increase, the balance of the foundation continues to be maintained both vertically and horizontally and there is no visible collapse, except for small rough settlements of the foundation.
- If the settlement is to be maintained, there is also a need for a continuous increase in vertical load.
- The soil outside the area remains virtually unchanged and no rotation occurs.
- This fault is typical in very loose sands or in soft or very soft cohesive soils.

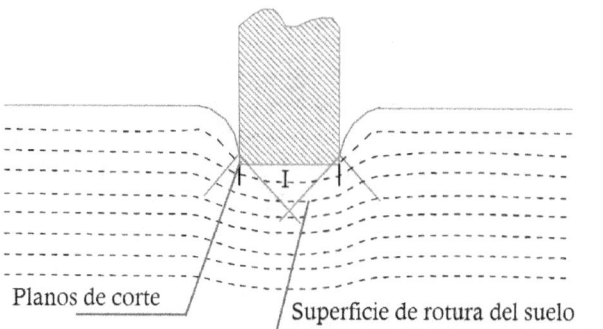

Figure 86. Punch fault scheme: I s elastic state wedge.

Figure 87. Settlement / Load by unit area q.

The ground fault surface will not extend to the ground surface. Beyond the ultimate failure load (qu). The graph load vs settlement, will be very steep and practically linear

SOLUTIONS TO FOUNDATION FAILURES

TERZAGHI SOLUTION

There are several known theoretical studies that can be applied in solving problems related to the load capacity of foundations in different soils. These theoretical studies include Prandtl, Fellenius and others. However, a less accurate but simpler solution to the problem is the one proposed by Dr. Karl Terzaghi and which has proven to be approximate enough for all cases in the field of practical application.

For running or continuous foundation:

$$q_d = cN_c + \gamma z N_q + 0.5\gamma B N_w$$

For square shoes and general cut

$$q_d = 1.3cN_c + \gamma z N_q + 0.4\gamma B N_w$$

For circular shoes and general cut

$$q_d = 1.3cN_c + \gamma z N_q + 0.6\gamma R N_w$$

$q_d \bullet$ foundation limit load capacity

SOILS AND FOUNDATIONS

No load capacity factors due to soil cohesion, overload and weight, respectively
(c) soil cohesion

γ • volumetric weight of the soil
z-depth of flattening
B- shoe width or smaller dimension
Circular shoe radius

Table load

$\emptyset^o=$ friction

TABLE: Terzaghi load capacity factors			
$\emptyset^o$	Nc	Nq	Nw
0	5.7	1	0
5	7.3	1.6	0.5
10	9.6	2.7	1.2
15	12.9	4.4	2.5
20	17.7	7.4	5.0
25	25.1	12.7	9.7
30	37.2	22.5	19.7
34	52.6	36.5	35.0
35	57.8	41.4	42.4
40	95.7	81.3	100.4
45	172.3	173.3	297.5
50	347.5	415.1	1153.2

13. Terzaghi capacity factors

Internal angle

$N_q = e^{\pi \tan \phi} \tan^2 \cdot \left(\frac{45+\phi}{2}\right)$ $N_c = c \cdot \cot \phi (N_q - 1)$ $N_w = 1.8 (N_q - 1) \tan \phi$

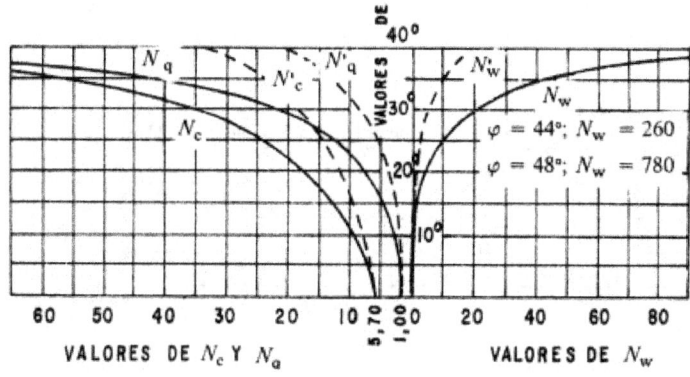

Figura 20.6 Factores de capacidad de carga para aplicación de la teoría de Terzaghi.
Figure 88. Graph Load Capacity Factors, Terzaghi.

SKEMPTON SOLUTION

Thinking in terms of fault surface length, a deeper foundation would have a more developing surface than a shallower one and therefore *soil cohesion would* work more. Skempton determined, experimentally and with some intuitive criteria, that the value of Nc *is* affected by the desplant depth of the foundation, growing, to some extent, as the flattening depth increases.

Skempton proposed that a Terzaghi-like expression be used to determine load capacity in purely cohesive soils:

$$q_d = cN_c + \gamma z$$

But with the difference that **nc no** longer has the fixed value of 5.7 for, but varies with the $\emptyset = 0z/B$ ratio at which z is the depth of dethinking the foundation and B *the* width of it.

The values proposed by skempton for N are those cited in the following table:

z/B	Nc	
	Circular or square shoe	Continuous shoe
0	6.2	5.14
0.25	6.7	5.60
0.60	7.1	5.90
0.75	7.4	6.20
1.00	7.7	6.40
1.6	8.1	6.80
2.00	8.4	7.00
2.50	8.6	7.20
3.00	8.8	7.40
4.00	9.0	7.50
>4.00	9.0	7.50

Table 14. Skempton Load Capacity Values.

CHAPTER 6:

TECHNICAL TERMS USED IN SOILS AND FOUNDATIONS

1. **Groundwater:** the one that can move in the saturation area of a mass of soil or rock by the effect of gravitational pull.

2. **Mechanical analysis:** granulometric analysis using sieves.

3. **Bolus:** fragment of rock rounded by abrasion or by meteorization whose diameter is greater than 30 cm.

4. **Pressure bulb: an** area limited by an isobara of stress, arbitrarily selected in a mass of soil, or rock subjected to a load.

5. **Quarry:** excavation on the surface of the earth for the exploitation of minerals or building materials.

6. **Ultimate support load capacity, qu: maximum contact effort that can withstand a soil mass without a catastrophic settlement of the foundation and the structure supported** by it, which corresponds to the theoretical value of the maximum support capacity of a soil. In the calculation of support capacity, account should be taken of soil shear resistance at foundation level and settlements likely by consolidation effect.

7. **Casagrande Plasticity Chart: a graph depicting the values of the liquid boundary and the plasticity index in a** Cartesian coordinate field to discriminate against clays and alms according to the relationship between those values. This chart is systematically used in unified soil classification.

8. **CBR:** California acronym Bearing Ratio. California Support Relationship. Measurement of a soil's relative resistance to penetration under controlled conditions of density and moisture content. It is the ratio of the effort required to penetrate a given material to the stress required to penetrate a reference material (pavement-based crushed rock) whose penetration resistance under normalized conditions is known.

9. **Density, s:** Mass of a body or material per unit volume. Numerical

relationship between (a) the mass and (b) the volume of a body.

10. Elasticity: ownership of materials that deform proportionally to the efforts to which they are subjected and recover their original shape and dimensions when the application of such efforts ceases.

CONCLUSION

It is of great importance to know about the Soils and Foundations since they have an important influence on the consolidation of any Constructionproject directly and indirectly, during the development of the evaluation it was possible to know more about the concepts and importance of soils and foundations in the development of a project, bases and criteria are acquired to be able to define and address situations of daily life where the question on what type of foundation to use and what type of soil or terrain will be answered.

Recommendations

It is recommended to always take into account before starting any construction project to carry out the relevant soil study to determine land characteristics and foundation and structure recommendations. which will be of use to us on certain occasions.

Bibliography

Villalaz, C.C. (2016). *MECCANICA OF SOILS AND FOUNDATIONS. (6 th ed.)* . by Editorial Limusa S.A. de C.V., Grupo Noriega Editores. C.P. 06040, Balderas 95 Mexico City.

Constructora Melendez S.A (2015) . *SOIL STUDY PROJECT VIS AMBAR MANZANA 2F CITY MELENDEZ.* SANTIAGO DE CALI, CPS. By Carlos H. Parra & Asoc. Civil engineers.

Botia Diaz, W.A. (2015). *MANUAL OF SOIL TESTING PROCEDURES AND CALCULATION MEMORY.* New Granada Military University, Engineering Faculty, Engineering Program. Bogota D.C.

Tomás, R., Santamarta, J.C., Cano, M., Hernández-Gutiérrez, L.E., García-Barba,J.(2013). *GEOTECHNICAL SOIL AND ROCK ASSAYS. UNIVERSITIES OF ALICANTE AND LA LAGUNA.* . . http://web.ua.es/es/ginter/ or http://ocw.ull.es/(date of access). License: Creative Commons BY-NC-SA.

Luis, G.V., (2011). *DESIGN OF FOUNDATION STRUCTURES ACCORDING TO NSR-10. NATIONAL UNIVERSITY OF COLOMBIA HEADQUARTERS MEDELLIN.* Race 80 No. 65-223 Block M1-107

NSR-10. (2011). *COLOMBIAN REGULATION OF CONSTRUCTION SISMO RESISTANT.* Bogota, Colombia.

Geovany Andrés, M., (2017). *ANALYSIS AND DESIGN OF STRUCTURES. Class Annotations.* Cali, Colombia: Universidad del Valle. Faculty of Engineering.

Porfirio, S.R., (2012). *PRESSURE DISTRIBUTION AND LOAD CAPACITY. BAJA CALIFORNIA AUTONOMA UNIVERSITY.* Faculty of Engineering, Architecture and Design Ensenada Unit.

Andersson, R.M., (2017). *SOILS AND FOUNDATIONS. Class and Tutoring Annotations.* Cali, Colombia: Universidad Santo Tomas de Aquino "USTA" CAU Cali, VUAD "Vice-Chancellor of Open and Remote University". Faculty of Science and Technology.

Fredy Andrés, M.R., (2017). *SOILS AND FOUNDATIONS. Technical Concepts.* Cali, Colombia: Universidad Santo Tomas de Aquino "USTA" CAU Cali, VUAD "Vice-Chancellor of Universidad Abierta y a Distancia". Faculty of Science and Technology.

Fabian, H.P., (2012). *GEOTECNIA BASIC DICTIONARY. Editorial rights © Fabián Hoyos Patiño 2001, 2005, 2012,* Medellin, Colombia.

ABOUT THE AUTHOR

Andersson Rincón Molina, SENA Construction Technologist (2013), SENA Works Supervision Specialist (2016-Currently), Professional in Construction, Architecture and Engineering Universidad Santo Tomas (Currently).